Work Organization

A Study of Manual Work and Mass Production

Ray Wild

Administrative Staff College, Henley, England

A Wiley–Interscience Publication

JOHN WILEY & SONS

London · New York · Sydney · Toronto

Library of Congress Cataloging in Publication Data:

Wild, Ray.
Work organization.

'A Wiley–Interscience publication.'
1. Work design.   2. Assembly-line methods. I. Title.

T60.8.W54   1975   621.7'5   74–13085
ISBN 0 471 94406 8

Printed in Great Britain by J. W. Arrowsmith Ltd.,
Winterstoke Road, Bristol, England.

Work Organization

Preface

The contents of this book are based upon a part of a report of a study supported by a Whitworth Fellowship Award, and undertaken during 1972 and 1973. I am indebted to the Whitworth Committee of the Science Research Council for permission to publish this material in this way and am also grateful for the guidance that they provided before and during my study and their comments on my draft report. I am grateful also to the numerous companies which provided facilities and to the many people who provided their time. Amongst my colleagues I would single out Mr. C. Carnall and Mr. D. Birchall for their assistance in the preparation of some of the material in this book.

I hope that this book will be of some practical value to the reader. I make no claim to have discovered anything novel, and have attempted simply to set down some of the things which I have found, and which have been of assistance to me in this area.

RAY WILD
Henley-on-Thames, 1973

Contents

CHAPTER 1

Objectives and Scope

The primary objective in this book is to examine present practice, current trends and developments, and to identify possible future practice in certain aspects of mass production. Two criteria will be taken into account—the operational efficiency of the production system, and the behavioural or human implications of the system. Thus we shall be concerned with manual work in mass production, the influence of the production system on the worker and vice versa. The ultimate aim is to attempt to formulate principles for the design of such systems and the design of such mass-production jobs, to satisfy both operational and behavioural considerations.

The focus will be the mass production of *discrete items* such as motor vehicles, domestic appliances, etc., that is, *complex* discrete items rather than nuts and bolts, etc. The emphasis will be upon multi-operation systems where the amount of manual work is significant rather than the type of mass-production systems found in the process industries such as petrochemicals. The emphasis will be upon assembly work. However, it will be argued that the views developed from the examination of such work are relevant in other aspects of mass production. Indeed it might be argued that the subject matter of the latter part of this book is of relevance in most areas where repetitive work abounds, whether this be 'blue-collar' or 'white-collar' work. In other words, the wider area of concern embraces the jobs and the work of individuals and the work systems which determine, influence or constrain those jobs. It is, nevertheless, convenient to take as the point of entry one particular class of operating system. Such an approach is appropriate since much of the information which will be of relevance relates to manual mass-production work. Such situations are widespread and much of the recent popular interest in the subject has focused on mass-production work, particularly assembly work.

The subject matter embraces many disciplines and functional/professional areas as well as being a topic of widespread current interest. The amount of public attention to the topic appears to be increasing; in fact the recent popular (indeed almost euphoric) interest has given rise to widespread discussions of topics such as the 'quality of working life' and 'blue-collar blues'. Similar explosions of interest have occurred previously and will doubtless recur, for this is a subject area in which the causes of concern are unlikely entirely to disappear.

1

The approach adopted in this book will be entirely descriptive. An attempt will be made to construct a body of knowledge rather than to test specific hypotheses. Of necessity, a wide range of topics will be covered since breadth of coverage is necessary if an adequate understanding is to be developed. No substantial claims are made for the originality of the views presented and every effort has been made to avoid esoterics. In fact, on reflection, the message of this book is very simple. However, we take the view that it is not only the end-product which is important but also the presentation of the necessary background information and the identification of the route to this end.

This will not be the first time that our subject matter has been examined, nor will it be the last; indeed some of the topics to be covered (e.g. 'job enrichment') have become, and may remain, major literary issues. However, in some instances such exposures have served only to confuse the fundamental issues involved. It seems that despite the previous work in the area a considerable amount of misunderstanding and misconception is evident. Certain important issues appear to have been missed or avoided, and an evangelic approach has in some cases evolved which attempts to be persuasive rather than informative. Furthermore, it would seem that in many cases a rather narrow approach has been employed, often concentrating exclusively on behavioural issues. Such observations led to the adoption of what will hopefully be found to be a realistic, comprehensive and balanced treatment.

This material is presented in six main parts, the order of these parts and their contents being a reflection of the manner in which this study developed. A survey/case-study type approach is adopted throughout. Part 1 covers some of the background, whilst in Part 2 a body of knowledge or information base is developed. In Part 3 case material is presented and examined in order to test the models developed in Part 2. Through Parts 4 and 5 certain principles are identified prior to their combination and presentation in Part 6.

Part 1
Background

An examination of the types of system employed for the mass production of complex discrete items, a survey of current developments affecting such systems and the benefit and criticisms of the principal system employed.*

* Readers familiar with the technology of mass production may wish to omit Chapter 2.

CHAPTER 2

Mass-production Systems

Before attempting to identify the recent and current changes taking place in this field, it is necessary to specify terminology and develop a means to classify and to distinguish between systems employed for discrete-item mass production. The identification of the basic types of mass-production system, some understanding of their relationships one with another and their characteristics, is a prerequisite for later chapters. Equally, however, any subsequent examination of mass production may produce information which could assist in developing an adequate classification system. For this reason the remainder of this chapter can only be devoted to the *tentative* identification of what appear to be the basic systems appropriate for the mass production of complex discrete items. Subsequent observations may lead to the modification of these views.

MASS PRODUCTION

Mass production is generally employed as a generic rather than specific term.[1] Although the term appears to have originated in the early 1900s, the types of systems now described are considerably older. Large-quantity production is as old as large-quantity demand, hence the concept is not new, only perhaps the manner in which the concept is translated into practice. Large-scale manufacture of certain items (e.g. bricks) began several centuries B.C.; however, the production methods adopted then differ from those that might be adopted today in that such 'mass production' was generally achieved through the use of a massive labour force. It was not therefore mass production in the modern sense of the term, since large output was achieved by the multiplication of effort rather than by the adoption of different production principles.

One stimulus for mass production was the invention and increasing availability of mechanized methods of production.[2] Tools such as lathes, drilling machines, forges, etc., were important, their development giving rise to perhaps the simplest aspect of mass production, namely the quantity production of single-piece items from single machines. A second aspect of mass production deals with the manufacture of more complex items, such as domestic appliances, motor vehicles, etc. Such items, because of their complexity or composite nature, cannot usually be manufactured by one tool or piece of equipment. They normally require the services of several facilities. The efficient mass production of such items, a more recent development than quantity production, is dependent therefore on the use of the flow principle, i.e. the continuous or near-continuous flow of the products through or past a series of production facilities. *Flow production* is most easily achieved for products which flow naturally. For example, in petroleum refining, the product and the raw material have a propensity to flow, and the design of the flow process is facilitated. In contrast, 'hard' discrete items such as engine cylinder blocks, motor vehicles and domestic appliances do not possess this characteristic. Hence considerable effort must be made to design flow systems for their manufacture.

The mass production of complex discrete items using the flow principle is one of the most important achievements in manufacturing technology and one of the most important aspects of mass production. Indeed, the importance of this method of manufacture is such that, for many people, the term mass production is synonymous with flow-line production.

On reflection, it appears that the generic term 'mass production' embraces two technologies—Exhibit 2.1.[1] Mass production and flow production are not

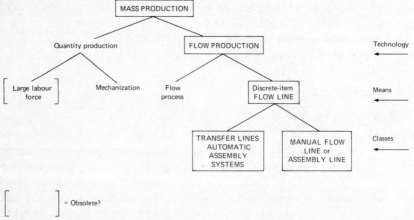

Exhibit 2.1 Mass-production systems

necessarily the same, since mass production only gives rise to flow production when necessitated by the nature of the product. Flow production consists basically of two subsections, namely flow processes designed for the manufacture of large quantities of bulk, fluid or semi-fluid products, and flow lines, which use the same principle of efficient material and product flow in the manufacture of large quantities of complex, discrete items.

Flow lines in which manual labour is used essentially for product assembly are often referred to as *manual flow lines* or *assembly lines*, whilst those using automatic material transfer between automatic machining 'stations' are normally referred to as *transfer lines* or *machines* or, in the case of assembly work, *automatic or mechanized assembly systems*.

Quantity production through mechanization is common in the engineering industry. The use of discrete-item flow lines is a comparatively recent innovation in engineering manufacture, deriving from the development of quantity production and the use of interchangeable parts, both of which in their turn depended to a large extent on the mechanization of production processes. Since we are concerned primarily with the mass production of complex and discrete items we will be concerned largely with flow production by means of discrete-item flow lines and hence with manual flow or assembly lines, transfer and automatic-assembly systems, and with the derivatives of, or alternatives to, such systems.

SYSTEMS FOR MASS PRODUCTION OF COMPLEX DISCRETE ITEMS

Manual flow lines

Manual flow lines are used predominantly for the assembly of items from a number of separate parts. They undoubtedly provide the principal system for such production.* They are characteristically labour-intensive systems, although some degree of process or handling mechanization may be employed. A major factor affecting the design and performance of such lines is the variable time required by workers at stations to carry out their repetitive tasks.

Manual flow lines consist essentially of a series of related work stations. Each station is manned by one or more workers and the total work content of the item is divided amongst the stations in order to obtain a line balance.

By identifying the basic types of line frequently used in industry, the following classification can be derived:

Moving-belt lines in which items are carried along the full length of the line on a moving conveyor belt, etc., items either:

(1) Remaining on the line during assembly (e.g. motor-vehicle assembly), or
(2) Being removed at stations, processed and replaced on the line.

'Non-mechanical' lines consisting of a series queue of stations. Items are transferred between stations either manually or by means of a conveyor, but cannot be carried past a station without processing, hence queues of items may form before stations on the line.

Space for buffer stocks of partially completed items will normally be provided between stations on non-mechanical lines. Such stocks serve to reduce the interdependence of stations and therefore to reduce the possible idle time occurring at stations due either to 'blocking', i.e. the inability of a

---

* The results of a survey of the nature and extent of the use of the systems identified in this chapter can be found in Wild, R., 'Mass Production in Engineering', *International Journal of Production Research*, **12**, 5 (1970), pp. 533–545.

worker to pass on a completed item to the next station, or 'starving', i.e. the lack of an item when required from the previous station. Workers on non-mechanical lines are not normally subjected to a mechanical-pacing effect, in that items may be retained at stations until completed. In contrast moving-belt lines provide a mechanical-pacing effect since items may pass from the station incomplete (fixed-item lines) or may pass by the station whilst a previous item is still being worked upon (removable-item lines).

The operating characteristics of moving-belt, fixed-item lines depend to some extent on the arrangement of work stations. Workers may be able to move up or down the line, and at any one time one or more items may be within reach of the worker. Such features reduce the mechanical-pacing effect and allow the worker to vary his work times slightly, rather than having to conform continually to the cycle time of the line.

A third method of line operation may be identified. Whilst it can be considered to be one type of fixed-item, moving-belt line* for our purposes it can also be classified separately as an *indexing line*, in which all items, normally fixed to the line, index or are transferred between stations either at a fixed interval (machine-paced) or on completion of the work cycle at all stations (operator-paced). In the former case a mechanical-pacing situation applies since the maximum time for completion of tasks is fixed by the operation of the line.

For all types of line, and for all methods of operation certain variations may exist, in particular more than one worker may be engaged at a station and stations may be 'paralleled' or duplicated. Additionally lines may be employed for the assembly of a single item, or two or more items simultaneously or in batches (mixed or multi-model lines respectively). Even though some types of line are free from a mechanical pacing effect, in practice a form of 'operator pacing' will normally exist, since workers on lines will be aware that their work-pace must correspond closely to that of their colleagues if a build up of items, or an excessive idle time, is to be avoided. The amount of idle time occurring at stations, the amount of work in progress and (for mechanically paced lines) the proportion of incomplete items produced are all sources of inefficiency in manual flow-line work.

Non-mechanical flow lines probably find greatest application in the assembly of relatively small items, which are easily handled and stored (for example clocks[3]). Moving-belt lines with fixed items are characteristic of much motor-vehicle assembly[4] and on such lines work stations are often manned by several workers and provision is normally made for workers to move out of their station, often by riding along on the line and then walking back to the next item. Moving-belt, removable-item lines are appropriate for the assembly of small, easily handled items such as instruments. Occasionally on such lines workers build up stocks off the line before and/or after their station, which has the benefit of reducing the pacing effect.[6] Indexing lines are infrequently found in

* Fixed-item, moving-belt line, with no station overlap and only one item available at each station.

labour-intensive flow-line work but may be adopted in the assembly of heavy items when some part of the process requires the items to be precisely located and stationary during processing.[7]

Individual and group assembly

The terms 'individual', and to a lesser extent 'group', work are often employed to describe labour-intensive assembly configurations in which there is an absence, or at least a relaxation, of the principal characteristics of flow-line work, i.e. intensive work flow and minimum work in progress. Such an approach is also often considered to provide the opportunity for the mass production of items without the necessity for the highly rationalized, paced and constrained tasks, characteristic of conventional flow-line systems.

Many assembly workers operate essentially independently. Often such workers are engaged on assembling large quantities of similar or identical items for long periods of time and, since such items might equally well be produced on an assembly line, these operators can be considered as being engaged in a form of 'individual mass production'. Typically parts would be brought to them in batches, and finished assemblies removed in batches; therefore, as far as the assembly operation is concerned, the system can be considered to be essentially non-flow. Such a system could be likened to the system of 'quantity production' referred to earlier. Group assembly work can be seen as a similar method of assembly where the item if sufficiently large may be built in a fixed position. Alternatively, workers may each be responsible for one task, as on an assembly line, but work virtually independently of one another by virtue of the large stocks of items accumulated between them. Such an arrangement might differ little, conceptually, from a flow line but efficient and smooth work flow might not normally be a dominant factor and furthermore, if workers are flexible, work balance is not a necessity.

Automatic or mechanized assembly

We will look more closely at the meaning of the terms mechanization and automation in Chapter 4; however, for our present purposes, automatic or mechanized assembly will be taken as referring to a method of assembling a product solely or largely by mechanical means without major manual intervention. In such cases, the operator's role will probably be confined to that of monitoring the system and making sure that component parts are available, although routine manual maintenance and tool changing may still have to be carried out. It is often necessary for an operator to orientate a part into the feeding mechanism, and on some machines there may be stations where an operator carries out an assembly or inspection task manually. Such systems, whilst still machine-intensive, are often referred to as semi-automatic or semi-mechanized, the distinction between mechanized and manual assembly systems in such cases being dependent upon the degree of system dependence on manual and mechanized elements.

Although assembly machines are still normally purpose-built, three basic configurations can be identified.

Single-station machines (fixed position)

On these machines the item is assembled in one position, usually on a jig or fixture. There is therefore no transfer or item flow. Such machines are limited in the number of parts they can assemble due to limited space, and are particularly useful when an assembly consists of several identical components, e.g. a number of washers on a spindle.[8]

Rotary or dial type

These machines can be of two types: continuous or intermittent movement, the latter usually being referred to as a rotary indexing type. Rotary machines have the advantage over the fixed-position type in that several work stations can be arranged around the periphery of the rotating table and so a greater number of components can be assembled. Items being assembled are indexed between stations, operations being performed or components added at each station. Operations are performed simultaneously at all stations and usually a completed item is produced for every index of the machine. The number of work stations is usually limited to five or six and machines tend to be compact. The majority of items assembled are of small to medium size due to the limited load capabilities of rotary tables.[9]

In-line type

In such machines items, usually positioned on a jig or fixture, are indexed in a linear fashion from station to station. These machines are the most flexible of the three basic types both in terms of number of stations and size of items assembled.

Two methods of operation are appropriate for multi-station assembly machines, i.e.

Synchronous transfer in which all items are transferred or indexed between stations at the same time. Machine-paced synchronous systems index items at fixed intervals, whilst in operator-paced systems transfer is controlled by workers at stations or the machine operator. The latter arrangement is often necessary when some amount of manual work is undertaken.[10]

Non-synchronous transfer involves the provision for buffer stocks of items between stations, since indexing of items at stations may take place at different times. Such a method of operation is beneficial when work stations with variable operations times, e.g. manual stations, are incorporated in the system.

Transfer machining

Whilst our primary concern is with the manufacture of complex discrete items, and thus with assembly work, it may be useful to look briefly at systems employed for the mass production of complex machined parts, i.e. transfer lines or machines. Examination of the types of machine in common use provides the following classification.

Rotary-drum type

Sometimes called the 'Trunnion Type', this type of machine is intended for medium production rates. The components are attached to fixtures on a rotary table or drum, operations being performed simultaneously on all the components on the machine. Operations can usually be performed on up to seven faces by machining heads positioned around the drum.

Rotary-indexing-table type

The machining units are *either* mounted on a column in the centre of the rotating table, or outside the periphery of the table, or both. As with the drum-type machine, loading and unloading is carried out during the machining cycle. The number of stations is usually limited as the space available for positioning machining heads around the table is limited.

In-line type

In-line machines operate on the same principle as the rotary machines, except that components are indexed linearly from station to station and machined simultaneously at each station. Two types of component transfer may be employed on in-line machines, i.e. Pallet Transfer, in which the components are clamped in position on a pallet, indexed from station to station and accurately located and clamped for the various operations, and Free Transfer, in which parts, not attached to pallets are indexed from station to station by lift and carry devices, for example, walking beam.

Link lines

A distinction is often made between 'transfer flow lines' and so called 'link lines', the latter consisting of a number of 'standard', often automatic, machines, interconnected by mechanical handling and conveying equipment.

SUMMARY

The basic systems, types and methods of operation identified above are compared and summarized in Exhibit 2.2. The similarities between transfer machines and mechanized assembly machines are readily obvious, as is the similarity, ignoring work-time variability, between indexing manual-flow lines and both classes of machine-intensive system.

MANUAL FLOW LINES		INDIVIDUAL AND GROUP ASSEMBLY		AUTOMATIC OR MECHANIZED ASSEMBLY		TRANSFER MACHINING		Characteristics		
Type	Operation	Type	Operation	Type	Operation	Type	Operation	Work flow	Mechanical pacing	Buffer stocks
		Individual		Single Station				Not intensive, irregular	X	X
		Group	Organized Self-organizing					Not intensive, irregular	X	✓
Non-mechanical				In-line	Non-synchronous	In-line	Non-synchronous	Intensive, irregular	X	✓
						Link lines		Intensive, irregular	X	✓
Moving belt	Fixed item							Intensive, regular	✓	Not normally
	Removable item									
	Indexing (machine and operator-paced)			In-line	Synchronous (machine and operator-paced)	In-line	Synchronous			
	Rotary					Rotary table		Intensive, virtually regular	✓	Not normally
						Rotary drum				

Exhibit 2.2 Comparison and classification of systems

The classificatory system employed above ignores so-called integrated systems, in which assembly and machining work are combined[11] and avoids the problem of defining degrees of automation. Such detail, whilst obviously important, is not necessary for our purposes, nor does it appear to be necessary at this stage for us to look more closely at the characteristics of systems and the manner in which these might be influenced in system design. Such considerations may well be relevant at a later stage, but the details above should be sufficient for our immediate, largely descriptive, needs.

Although sometimes included in the general class of flow-line processes[12,13] the cell system of group technology has not been included in this classification, since we are concerned solely with mass rather than batch production. However, since cell production can be seen as a system for the mass production of parts for batch-produced products, and since certain cell configurations bear a close resemblance to mixed or multi-model, non-mechanical flow lines, it may be necessary to pursue such similarities during subsequent sections.

References

1. Wild, R., *Mass-Production Management*, Wiley, 1972.

2. Ford, H., 'Forming & shaping operations for iron & steel', *Chartered Mechanical Engineer*, **20**, 4 (1973), pp. 53–58.

3. Reutebuch, R., 'Assembly lines for clocks and watches', *Sub-Assembly* (April 1972), pp. 28–30.

4. Murrell, K., 'Ford Escort production line at Halewood', *Sheet Metal Industries* (April 1968), pp. 237–269.

6. Franks, I. T., Gillies, G. J., and Sury, R. J., 'Buffer stocks in conveyor based work', *Work Study & Management Services* (February 1969), pp. 78–82.

7. Astrop, A. W., 'Aspects of BLMC Cofton Hackett plant—1', *Machinery and Production Engineering* (18 March 1970), pp. 402–406.

8. Cappuccio, E., 'Mechanisation of manual operations in industrial assembly', in Ross (Ed), *Proceedings of 1st International Conference on Product Development and Manufacturing Technique*, McDonald, 1969.

9. Green, R. G., 'Assembly machine configuration', *Automation*, **18** (1971), pp. 41–47.

10. 'Making parts for Odhner calculating and accounting machines', *Machinery* (21 April 1965), pp. 844–853.

11. Patton, W. G., 'Automatic assembly takes in machinery & inspection', *Iron Age* (13 March 1969), pp. 75–76.

12. Edwards, G. A. B., *Readings in Group Technology*, Machinery, 1971.

13. Thornley, R. H., 'An introduction to group technology', Paper presented to Group Technology Course U.M.I.S.T., 4–6 July 1972.

CHAPTER 3

Characteristics and Developments

Manual flow, or assembly, lines doubtless provide the principal system of mass production for the type of goods with which we are concerned.

Our objective in this chapter will be to identify the characteristics of this system for mass production, the development which might affect the nature and use of the system and the reasons and motives for such development. In other words—why might such development be appropriate? What is wrong with the conventional assembly line? In looking briefly at the characteristics of manual flow lines and flow-line work, we will attempt to outline some of the arguments advanced together with the recently* cited incidents and developments either in support or in opposition of assembly lines.

CHARACTERISTICS AND CRITICISMS OF ASSEMBLY-LINE WORK

The mass production of complex discrete items is currently characterized by the widespread use of flow-line systems, whether essentially manual lines, as in much assembly work, or machine-intensive systems, such as transfer lines and some automatic assembly machines. Manual flow lines, or assembly lines, have been in widespread use in industry for over fifty years, and some industries, notably motor-vehicle manufacture, have been and still are heavily dependent upon this system of mass production. The principles upon which such systems are based were well developed and proven some time before assembly lines themselves were widely adopted. For example, the use of interchangeable parts in assembly was proposed in the early 1700s, and used successfully around 1800. The division or rationalization of operations was in widespread use in industry by the mid-1800s, whilst conveyor systems had been designed and used by 1785. The widespread use of flow-line systems did not therefore represent a new departure in manufacturing technology, but rather the application and refinement of existing technology. Nor has there been radical innovation in the intervening fifty years, but rather a continuation of developments, e.g. the further rationalization of work on assembly lines and the development and introduction of further degrees of automation and mechanization. The development of machine-intensive, flow-line systems, e.g. transfer lines, has proceeded largely unchallenged, whilst the recent development and

* (1972–1973).

14

application of labour-intensive, assembly-line systems appears to have begun to meet opposition.[1,2] The basis for this 'opposition' is again comparatively well established, the characteristics of work on manual flow lines having been questioned for decades.[3] However, it is only in recent years that opposition has begun to increase to the extent of influencing system design. It might seem, therefore, that in some respects the development processes evident since the turn of the century are about to terminate. Recent publicity and debate, together with certain new developments in the mass-production industries might all be taken to indicate that a stage of discontinuity has now been reached, at least as far as manual flow or assembly lines are concerned.

Application of the principle of efficient and regular material flow in flow-line work leads to the minimization of work in progress and item throughput times, as well as the minimization of space requirements, simplicity of production control and reduction of indirect work. Such characteristics, however, also mean that flow-line systems are susceptible to disruption through the malfunction of the process; they are inherently inflexible and require accurate specification of tasks to permit efficient operation. Although not strictly a principle of flow-line production the division or specialization of the tasks performed on assembly lines is an important characteristic,[4] which facilitates line design. Work specialization facilitates learning, permits the employment of unskilled and semi-skilled labour, enables high performance levels to be achieved without undue physical or mental stress[5] and perhaps minimizes supervision requirements.

Flow-line production systems have contributed significantly to increased productivity and many of the characteristics of these systems have served as targets in the development of systems in, for example, batch production.[6] The introduction of manual flow lines in the manufacturing industry was heralded as a major development in the automation of production and the use of such systems clearly moved manual-assembly work significantly closer to the type of continuous-flow process normally associated with automated production. Other than those features associated with inflexibility and susceptibility to disruption the disadvantages of manual-assembly lines are less readily proven. Generally they relate to the nature of the work required on manual flow lines and the effects of such work on the workers. It is not generally agreed that all assembly-line workers experience feelings of boredom or monotony[7] or that such effects if they exist necessarily give rise to any form of counter-productive behaviour.[8] Nevertheless such possible effects of flow-line work are now beginning to receive closer scrutiny.

In general, recent criticisms of assembly-line work have been based upon the following premises:[9,10,11]

(1) Educational changes

Increases in the educational attainments of workers and the changing nature of their education are considered to be trends in opposition to the rationalization and constriction of manual work evident in assembly-line work. Today's school leavers are said to be more inquisitive, socially aware and knowledgeable than their predecessors, and are thus less likely to accept or to remain in the rationalized and repetitive industrial jobs typically occupied by their predecessors. Contemporary, extended secondary education is said to place more emphasis on learning, as opposed to reception, to develop social relationships and yet emphasize independence, whilst the availability of continuing formal and informal education is said to help develop awareness of social, political and commercial issues, which affect attitudes to industrial employment.

(2) Labour force and labour market changes

Comparatively low unemployment, increased mobility of labour, the expansion of the service industries, increased social security, including sickness and unemployment benefits, the availability of retraining schemes and the changing constitution of the labour force are given as reasons for the need to re-assess the nature of certain types of manual industrial work. The ready availability of alternative employment in many countries is said to encourage labour turnover. Absenteeism is thought to be indirectly encouraged by increasing wage levels and trends in the conditions of employment giving some workers 'staff-like' conditions for payment during sickness, etc. The replacement of piecework by time or day-work payment systems is also considered to encourage casual absenteeism. The employment of an increasing proportion of women, including married women, is thought to lead to increased labour turnover and absenteeism, whilst increased wages for women workers may substantially increase the cost of certain items assembled on manual flow lines. Immigrant saturation in countries may limit the supply of labour and promote revaluation of the nature of mass-production work.

(3) Social and political changes

Increasing public and official concern for the natural environment, amenities and the pollutant hazards of certain industrial activities is considered to be focusing public opinion on questions related to the quality of life and the social and environmental responsibilities of industry. Such concern is considered to embrace the quality of working life and the working environment, hence matters relating to the influence of technology on the nature of work, the workplace and the purpose of manual work are of increasing public concern. Personal freedom and responsibility at work are thought to be topics of increasing Trade-Union concern, negotiations over pay and conditions being supplemented, if not replaced, by bargaining on topics such as the elimination of boredom and monotony of work. Political and ideological pressures towards industrial democracy, worker participation and co-partnership will, it is argued, lead directly to pressures for the democratization of the 'shop floor'—such democratization being associated with worker autonomy, responsibility and decision making. The international mobility of labour, growth of international companies and Unions, and international political confederation are all likely to accelerate such developments.

(4) Changes in products and markets

Trends towards increasing product diversification, shorter product life and increased variety may increase the need for flexibility in the production system. Such flexibility, perhaps also sought to combat high labour absenteeism, may affect flow-line design. Increasing competition may focus attention on product quality which may also lead to modification in assembly-system design, whilst increased rates of demand changes and product and process innovation may necessitate reorganization of decision-making procedures at the 'shop-floor' level to provide quicker and more effective response and less resistance to necessary change.

The effects of boredom, monotony and alienation said to result from assembly-line working allegedly include industrial disputes, increased absence and labour turnover, and reduced quality and productivity. Recent industrial disputes at the Batignolles, Ferodo and Moulinex plants in France have been reported as being related to workers' attitudes to mass-production work. The five-week strike by assembly-line workers at the Renault Le Mans plant is said to have 'sparked-off' extensive discussions finally prompting the CNPF (the French employers' federation) to conduct a world-wide study culminating in a report on 'The Problem of the Assembly Line Worker'.[12] Similar strikes in America have been widely reported, for example the action in 1963 involving thousands of Detroit workers who stayed on strike following their Union's negotiation of an agreement which omitted provision for rest periods, reduction and control of line speeds.[12] Assembly-line workers at Olivetti in Italy are reported to have struck following the modification of their flow line to a system which increased the degree of work pacing, whilst Fiat in Italy have also experienced strikers by workers citing their dislike of their 'repetitive tasks'.[12]

The motor industry, popularly considered to epitomize modern mass production, has been the focus of much of the recent criticism of assembly-line work. Labour absenteeism at General Motors and Ford in the U.S.A. is reported to have doubled over the past ten years, with the sharpest climb in recent years.[2] On average 5 per cent of G.M.'s hourly staff are absent without explanation each day, whilst the rate on Mondays and Fridays is as high as 10 per cent.[2] Labour turnover at Ford has increased to over 25 per cent—many workers are said to walk off their jobs in mid-shift,[2] whilst absenteeism has on occasions led to plant shut-downs, with consequent loss of output and pay. Absence and turnover rates are comparable, and even higher in other plants in the motor industry. Absence in one Chrysler plant averages 8 per cent to 9 per cent midweek and 15 per cent on Mondays and Fridays.[13] Higher absence rates, reaching 15 per cent in some plants, are reported in France and Italy, whilst turnover is said to have reached 100 per cent per annum in certain Fiat plants. Absenteeism at Fiat's plants is reported to be as high as 30 per cent on some days, whilst even on an average day some 12,000 or 13,000 workers, representing 14 per cent of available labour, fail to turn up.[14] Such absenteeism compares with an average of 5 per cent only a few years ago, and is considered to be one of the main factors limiting Fiat's growth and profitability. The fact that employment has increased by 26 per cent in the last four years, during which output has risen by only 10 per cent, is seen to result primarily from the increased absenteeism.[14]

Absence and turnover amongst American manual workers is higher for younger workers and the proportion of such workers in the labour force is increasing. In ten years time the number of people under 34 in employment in the U.S.A. will have increased by 46 per cent.[12] Of the 740,000 hourly-paid workers in the American motor industry 40 per cent are now under 35, whilst approximately one-third of such employees at Chrysler, G.M. and Ford are under 30. More than half of Chrysler's American work force have been there less than five years.[2]

Quality performance is considered to be related to worker attention and vigilance, in turn affected by motivation and job satisfaction. Several studies have suggested that workers attitudes to repetitive work affect the quality of the products resulting from such work but only recently have there been reports of apparently deliberate negligence and sabotage. Again reports relate to the motor industry and include accounts of scratched paintwork, screws left in brake drums, tool handles welded under bodywork, and cut upholstery.

THE END OF THE ASSEMBLY LINE?

Widespread discussion of issues, such as those cited above, has not only focused attention on the shortcomings of assembly lines, but has also led to speculation about the continued existence of this system of mass production. In this section we will attempt to summarize two principal themes which have recurred throughout recent discussion of the future of assembly lines.

The redesign of assembly-line work

Arguments such as the above have inevitably given rise to some speculation on the future of assembly-line systems. The E.E.C. in one document,[1] subsequently withdrawn, appeared to argue that assembly-line systems should be abolished. Others have argued similarly, and furthermore many recent developments appear to support not only the view that such systems should, but also can, be abolished. Speculation about the future of conventional assembly lines received considerable momentum in the early 1970s as a result of the very considerable publicity given to developments in Sweden in the traditional 'stronghold' of the flow line—the motor industry. These developments at Saab and Volvo, whilst not unique, nor indeed the only developments of their type to be taking place at that time, gave considerable impetus to the general controversy. It was recognized that the apparent benefits of flow-line systems in the motor industry were considerable, and that the obstacles to change were substantial, and yet two major companies in what was generally considered to be a technologically, socially and economically advanced country had chosen to seek alternative methods for the production of vehicles. Although the work of these two companies is now widely known it may be worthwhile providing some details here, in order to sketch in a significant aspect of the background of our study.[15-22]

The Saab experiments

The introduction of an experimental assembly method in January 1972 was part of a series of schemes employed largely to counteract labour turnover and absence problems. Turnover on Saab's main assembly line in 1971 was 34 per cent with absenteeisms of 20 per cent. One year previously turnover touched 100 per cent.

In the new method there are seven working areas each staffed by three fitters, each of these teams being responsible for assembling a *whole* engine—which involves fitting some ninety parts onto a unit. The teams make up their own minds how they do the work but the time given for each engine is thirty minutes. Each can assemble an entire engine in thirty minutes, or all three can work together, in which case they have ten minutes. One group normally acts as a training area for new employees.

Saab had full Union cooperation in the introduction of this system and the efficiency of the new method of assembly was found to be as good as an equivalent assembly-line system. The output rate was unaffected by the new system but an increased cost of 20,000 SKr–30,000 SKr per year is involved through increased space and tool requirements.

The company had also reorganized chassis-assembly work by changing individual operations, by rotating employees amongst jobs and by giving increased responsibility to groups of workers. All such approaches had one thing in common—i.e. collaboration between various occupation categories to improve productivity and increase job satisfaction. Labour turnover and absenteeism in the company declined and although some improvement may have resulted from national economic changes the company believe that their attempts to improve work methods are essential to their future profitability and viability.

The Volvo development

The experiments were prompted in part by high labour turnover in the company, which in 1969 reached 55 per cent whilst absenteeism was running at 20 per cent. Turnover in 1971 was 30 per cent and the company needed to recruit every year about one-third of its 36,000 Swedish work force just to keep going, and to have about one-seventh of the work force in reserve because of absenteeism. Turnover was found to be concentrated amongst newly employed workers, many of whom opted out at the end of, or even during, their training period. Since the company devoted at least 125 hours training to each worker this turnover was considered to be extremely costly. The experiments were undertaken to 'get stability of production and satisfied people', 'to get people more involved in their job', to combat the monotony of the flow line and to give the employee more incentive and satisfaction.

The company began conducting experiments in 1966, concentrating initially upon schemes by which workers were rotated amongst several jobs. Some assembly-line workers were allowed to remain longer with the vehicle they were assembling, up to one hour being spent on one vehicle compared to a previous cycle time of only a few minutes.

In the design of the new vehicle-assembly plant at Kalmar a radical new approach was adopted. The new plant substituted a series of small independent 'workshops' for the conventional single long flow line. Each has its own entrance, changing room and relaxation area, and the work of assembling cars is split up between teams of workers, each having responsibility for one section. Within each 15–25-man team the workers decide how the work is to be split up and how and when they will exchange jobs. Buffer stocks between work areas permit variation of work speed.

In some cases these Swedish experiments were seen as a prelude to the wider reorganization of flow-line working in the motor industry. It was argued that the problems faced by industry in Sweden were those that would be faced in other European countries in the near future, and thus that these developments were indicative of future developments elsewhere. Equally, however, it was argued that just as certain conditions necessitate change so also do peculiar conditions facilitate certain types of change which might not be appropriate elsewhere. The industrial relations situation and certain cultural factors in Sweden were therefore given as reasons for the adoption of such changes. Furthermore, it was pointed out that both Saab and Volvo have indicated that their new assembly methods might not be applicable for all, or even the majority of, work in their companies. Vehicle output at Volvo and Saab is substantially less than that of most other manufacturers. Total vehicle production in Sweden is only 11 per cent of production in the U.K. whilst Volvo's new Kalmar plant was designed for only 30,000 cars a year—less than 10 per cent of the capacity of Ford's Halewood plant.[23] Total 1971 Volvo car output at 214,000 compares with Ford's U.K. output of 700,000 per year and over one million per year from British Leyland.[24]

It has been suggested that to assemble a car in England without automation and without the use of present-day assembly-line systems would triple the cost of the vehicle. Even the most enthusiastic reports on recent innovations designed to improve the motivational content of work recognize that the complete abolition of flow lines in automobile production would be a long, complex, costly and perhaps crippling project; nevertheless, one interpretation of current evidence might be that the days of the assembly line are numbered if not at an end, a view which finds some support in reports of experiments in other industries.

Attempts to restructure manual work in other industries often involved the replacement of flow-line working with other forms of manual work. Job redesign exercises have been widely reported, particularly in companies engaged in the manufacture of small products such as domestic appliances, electronics, radios, etc.[25,26,27] To some degree efforts to redesign such work have been encouraged by the growth of productivity bargaining and thus it might be argued that the objectives have more often included increased flexibility, and reduction of demarcation, than increased worker job satisfaction.

During the last five years research workers have increasingly begun to advocate the creation of work groups in production situations. Many such work groups have been formed, usually on an experimental basis.[28,29] The advocacy of this approach has received added momentum from the work conducted in Saab and Volvo although it has been suggested that group or team work is of limited application in the motor industry.[30] 'Cellular' organization, a form of group working in batch production, is also increasingly advocated.[31,32] The principal advantages of cell manufacture have conventionally been identified with reduced work in progress, improved throughput and greater flexibility, and only recently has the social and motivational potential of this form of organization been identified. Indeed in some respects it appears that the independent advocacy of group working in both mass- and batch-production points to the emergence of a common philosophy in production which conflicts with the continued use of flow-line systems. However the precise nature and characteristics of such a philosophy have yet to be presented.

Mechanization and automation of assembly-line work

Increased mechanization and automation has also been seen as a solution to the problems of worker dissatisfaction on repetitive assembly-line work; indeed, the elimination of monotonous 'dehumanized' jobs is one widely claimed benefit of automation and mechanization. It has been argued that 'machines should be built for all jobs that are physically arduous or mentally stultifying' and that 'men should be reserved for tasks that are non-routine—those requiring versatility and intellectuality'.[33]

Recent surveys have shown that automation and mechanization is normally pursued in order to provide reduced labour costs, improve output quality, increase output and overcome problems of labour shortages and absenteeism.[34] The reduction of the working week, movement of employment from blue-collar to white-collar jobs, the miniaturization of parts and products and protection from labour disputes are given as further justification for increased automation, whilst inflexibility is often seen as a major disadvantage of highly mechanized or automated production. This disadvantage, however, is now beginning to yield to further innovation, in particular the development of industrial robots or programmable automatic-handling devices. There are currently thought to be approximately 1200 at work in industry, mainly in the U.S.A. and Japan, where applications involve die-casting, forging, plastic moulding, machine loading, welding, assembly and conveyor loading.[35]

Automation and mechanization is perhaps less developed in assembly work than elsewhere,[37] however recent years have seen considerable development, particularly in respect of small products, such as clocks and watches, automotive accessories,[39] etc., and electronics. Developments in the assembly of larger items, such as motor vehicles, have until recently primarily affected materials handling rather than processing; however, for high outputs a comparatively high degree of automation may now be practicable even in previously labour-intensive mass-production assembly work. The following case, for example, describes one recent approach to motor-vehicle body assembly—an aspect of vehicle production that conventionally requires considerable repetitive manual work.

Automation of car fabrication

The Vega car was introduced by Chevrolet in 1970, all production being located initially at a plant in Lordstown, Ohio. From conception the car was planned as a cheap (average price $2500) and high-output-quantity vehicle (target annual production 400,000). It was designed to be simple to assemble. The body comprises 1231 pieces, approximately one-third of that of other cars previously produced at the plant. All four models share many common parts, and large panels and subassemblies were used rather than smaller components. Fewer joints and seals are required than for a conventional car. The Vega body plant was the most automated vehicle plant in the world. Output rate was to be one hundred vehicles per hour.

Whilst conventionally approximately 80 per cent of the welds on a vehicle body are accomplished manually, 95 per cent of the 3900 Vega body welds are performed by seventy-five automatic welding devices including twenty-five industrial robots. Each Unimate robot has five-axis programmable control to work within a positional accuracy of 1/16 inch. The 578 different body parts (43 per cent fewer than normal) are designed to facilitate automatic welding by providing for open approach and single-plane motion for seam and spot welding. A typical Unimate engaged on spot welding is capable of automatically making 2000 welds per hour. Bodies on this welding line move on a 'lift and carry' principle, locating pins being provided to prevent relative movement between body and welding equipment. Other sections of the line move continuously, the equipment being powered to move with the line and then return to the next station. The intended cycle time of the lines was thirty-six seconds; however, as some travel time is required, work times for certain tools is as low as twenty seconds.

The number of necessary adjustments during assembly were minimized, hence doors and bonnet hinges are welded rather than bolted. Doors carried by hydraulic positioners move into position as the line moves and upper and lower hinge pins are inserted by machine. The bonnet and hinge pins are assembled similarly. Automatic spot welding on a two-stage transfer-type line is employed for part of the underbody structure. Initial body assembly is accomplished on a continuously moving line. An underbody is automatically loaded onto a building trolley from a storage bank, whilst the fixtures holding the side assemblies are automatically moved into position and locked accurately in place on the sides of the trolley. The appropriate complete roof structure is sequenced from store, automatically dropped and clamped into place.

Some welds are made manually and an automatic tack welder indexes to the moving body making sixty-five welds on each side in 11·5 seconds. Additional automatic welders complete the body, and the fixtures are automatically removed and the body travels to the indexing conveyor on which the robot devices are employed.

Self-contained memory systems monitor the function of each Unimate. If a machine fails to complete its cycle of operations a failure message is displaced, and malfunctions are relayed to a later rectification station. To minimize expensive off-line rectification the quality control programme attempts to identify unsatisfactory trends, e.g. one automatic inspection station checks the dimensional accuracy of an item by determining the relative location of location points, provides a digital read-out, calculates and signals adverse trends.

The plant when fully commissioned was heralded as a major development in production automation—'the pattern of vehicle production of the future'. The car and the production facility were designed for each other, to provide high line speed, high quality and low vehicle cost.

Although many examples exist of increased mechanization being accompanied by the elimination of tedious manual tasks, and the emphasis of skilled work,[47] it is equally clear that *in certain cases* such processes of technological development can adversely affect the nature of the remaining manual work.

One well-known example[48] relates to visual inspection of Coca-Cola bottles, which was found to be so tedious that occasional bottles of '7-up' had to be placed on the line to maintain vigilance. Whilst the potential of automation to eliminate boring jobs is evident, especially in certain types of production, e.g. the process industry,[49] it has been questioned whether this advantage necessarily ensues from the automation of assembly work either in the short or in the longer term. The following report, whilst not exclusively concerned with the effect of the automation of work, deals briefly with what has become the most publicized case of its type.[40-46]

The Lordstown disputes

Considerable production difficulties were experienced following the opening of the highly automated Lordstown Vega plant. The employers contended that some work slowdown and disruption had taken place, such that during December 1971 and January 1972 incomplete and damaged cars had been produced at such a rate that repair areas were full before the end of an eight-hour shift. Up to this time an estimated 12,000 cars and 4000 trucks had been lost at a value of $40 millions. By February 1972 over 5000 grievances had been lodged by workers—1000 dealing with overwork and work standards, whilst the Union contended that 700 jobs (from a total of approximately 8000) had been eliminated during reorganization and 'speed-up' in the plant. Management alleged deliberate neglect by production-line workers and even sabotage of cars and equipment. Broken windscreens and mirrors, cut upholstery, keys broken in locks, washers in carburettors and bent signal levers had been reported, whilst some items were said to pass through several stations on the lines without the necessary work being performed on them. Examples were given of the Unimate welding machines being reprogrammed to weld bodies in the wrong places and of finished cars having had all of their doors locked.

On February 2, 1972 the local U.A.W. voted to strike if no settlement was reached on the 5000 grievances then at issue. The strike began at 2 a.m. on March 4, following a failure of 'last-minute' negotiations. 7800 workers in the plant were involved whilst 8800 were affected at other plants. The plant was closed on March 6 and the strike lasted until March 26. This strike, together with the disruption immediately preceding it, were estimated to have led to a loss of $160 millions in production, or 70,000 vehicles, whilst payroll losses to striking workers were estimated at $11 millions.

The average age of workers in the plant was approximately twenty-four years, and unlike many car plants in America, few of the workers were coloured (as little as 100 from the total of nearly 7000) and only about 500 women were employed. Whilst company spokesmen contended that the dispute was unconnected with the discontent of workers on assembly-line work, much of the publicity and discussion accompanying the dispute focused on the supposed clash between the 'new' and young generation of auto workers, and the characteristics of traditional motor-industry work, in particular the nature of such work on a highly automated 100 vehicle per hour line. Following the 1970 strike G.M. and U.A.W. officials had issued a joint statement recognizing that younger workers were more likely to dislike their jobs in the industry—the company had begun to introduce counselling schemes, lengthened induction periods, etc., to improve the motivation of workers, having recognized that the young work force might be susceptible to high absenteeism, etc. Team working and job enlargement had been proposed by the U.A.W. during negotiations in 1970 and subsequently the Union had begun to press for a shorter working week, and flexible work-hours. Discussing monotony the head of the company assembly division (G.M.A.D.) said '. . . there is a great deal of misunderstanding about that (monotony) but it seems to me that we have our biggest problems when we disturb that "monotony"', whilst a training director in the company reportedly said '. . . it is not the repetition but the chaos of the assembly process that is most discouraging.'

The 1972 strike ended on March 26. The Union announced that all but 130 of the 700 workers who had been laid off since G.M.A.D. took over the running of the plant would be returning to their jobs, and that 800 of the 1200 outstanding grievances had been cleared and paid. Additionally some workers had won back-pay, and several changes and layout and work-assignment procedures had been agreed.

Widespread discussion of the dispute and grievances continued and was further encouraged by a subsequent strike at the plant in October 1972, and strikes at other G.M. plants, including one lasting 175 days at Norwood, Ohio. The Government centre for auto-safety began an investigation into possible violations of the safety laws through faulty workmanship and parts on the Vega, whilst considerable attention was given to the unreliability of production equipment and the effects of such failures when working on cycle times approaching thirty-six seconds.

During 1972 the company released several reports dealing with their efforts to combat unrest on assembly lines. A Vice-President of Personnel Development was appointed with responsibility for effective worker motivation, whilst a research group was established to study the problem. Surveys were conducted and organizational development programs begun whilst various exercises were introduced at specific plants. In the meantime the St. Therese plant at Quebec was being designed for production of Vegas with an output rate thought to be in the order of 50–60 vehicles per hour.

SUMMARY

The evidence available both for and against assembly lines appears to be somewhat inconclusive. Whilst the importance of this system in contemporary mass production is beyond dispute, it is not clear whether, or how, the present situation will persist.

Manual production flow lines presently provide the principal means for the mass production of complex discrete items, particularly assemblies, in most industrialized countries. Contemporary manual flow lines, or assembly lines, have developed over the past fifty years through the continued application of certain basic and simple principles; however, it might now appear that such development is drawing to a close. In fact some people have argued that the principles have been overdeveloped.

The main criticism of this system of mass production relates to the role of the worker. For many years it has been suggested that the work required by flow-line systems is no longer appropriate for the workers who 'man' such systems. Recently such criticism has increased to such a degree that terms such as 'blue-collar blues' and discussions of the 'end of the assembly line' are commonplace. The principal target for this debate has been the motor industry, which popularly epitomizes modern mass production. The high absenteeism and turnover rates, low quality, increasing industrial strife, and suggestions of violence and sabotage in motor-vehicle manufacture have been given as evidence of the need to rethink the production methods used in this industry. Experiments at Saab and Volvo in Sweden have supported the growing contention that the demise of the assembly line is at hand, and it has been suggested that it is now unlikely that any new motor plant will ever be designed along previous conventional lines.

The design of the Chevrolet Vega plant in America provided a complete contrast with the recent Swedish approach, since it was conceived as the world's most automated vehicle-production plant. The severe series of strikes and disputes which followed the opening of the plant is also seen in some respects as a warning of the effects of the further mechanization of flow-line work and as evidence of the need to rethink mass-production methods. However, it is equally clear that in many circumstances increased automation and mechanization is considered to provide the best solution to possible problems of boring repetitive work, through the elimination of such work.

If we adopt a normative view and take conventional assembly lines as our starting point then in this necessarily cursory review we have identified two dimensions of current development, associated with:

(1) The increasing mechanization and automation of work, i.e. mechanized assembly.
(2) The modification/redesign of the jobs of assembly-line workers, generally involving the provision of 'larger' and less constrained jobs.

These two approaches are evident in many industries and in some cases the different approaches are apparently being adopted in ostensibly similar situations. In the following chapters we shall attempt to shed some light on such seemingly opposing approaches, through an examination of recent changes in production systems, and an examination of the reasons for such changes. In the next two chapters we shall look at these two dimensions of development in some detail in order that we might build an information base from which to work.

References

1. 'Guidelines for a social action programme', *Commission of the European Communities*, COM(73), 520, Luxembourg (18 April 1973).
2. Gooding, J., 'Blue collar blues on the assembly line', *Fortune* (July 1970).
3. Friedmann, G., *The Anatomy of Work*, Translation published by Heinemann, 1961.
4. Woolard, F. G., *Principles of Mass and Flow Production*, Iliffe, 1956.
5. Wild, R., *Mass-Production Management*, Wiley, 1972.
6. Demyanyuk, F. S., *Technological Principles of Flow Line and Automated Production*, Vols. 1 and 2, Translation published by Pergamon, 1963.
7. Thackray, J., 'American industry's strange behaviour', *Management Today* (May 1973), pp. 84–89.
8. Vroom, V., *Work and Motivation*, Wiley, 1964.
9. 'An escape from the machine', *The Times*, London (24 April 1973).
10. Tromp, P., 'Man at work in an expanding world', *Occupational Psychology*, **45** (1971), pp. 173–181.

11. Trafford, J., 'Why autocracy must die', *Financial Times* (17 November 1972).
12. Bosquet, M., 'The "Prison Factory" ', *New Left Review*, **73** (May/June 1972), pp. 23–34.
13. Serrin, W., 'The assembly line', *Atlantic*.
14. Ensor, J., 'Absentees spike Fiat's growth', *Financial Times* (14 November 1972).
15. 'Saab abolishes the car assembly line', *Manufacturing Management* (June 1972).
16. Palmer, D., 'Saab axes the assembly line', *The Times* (23 May 1972), p. 17.
17. Chaote, R., 'Saab's "teamwork" success', *Financial Times* (18 May 1972).
18. Willatt, N., 'Volvo versus Ford', *Management Today* (January 1973), pp. 43–50.
19. Willatt, N., 'Can Volvo kill the assembly line?', *Financial Times* (3 October 1972).
20. Mounter, J., 'Volvo spending £20m on plant to eliminate mass assembly lines', *The Times* (20 June 1972).
21. McGarvey, P., 'Car making without the boredom', *Sunday Telegraph* (4 February 1973).
22. 'The will to work and some ways to increase it', *Life* (September 1972).
23. Murrell, K., 'Ford "Escort" production line at Halewood', *Sheet Metal Industry* (April 1968), pp. 237–239.
24. Harper, K., 'Shorter hours major issue in pay claim by 13 Ford unions', *Guardian* (17 January 1973).
25. Gooding, J., 'It pays to wake up the blue collar workers', *Fortune* (September 1970).
26. Myers, M. S., *Every Employee a Manager*, McGraw-Hill, 1970.
27. Maher, J. R., *New Perspectives in Job Enrichment*, Van Nostrand Reinhold, 1971.
28. Emery, F., Thorsrud, E., and Lange, K., *Field Experiments at Christiana Spigerverk*, Tavistock, London, 1966.
29. Argyle, M., 'Working in groups', *New Society* (26 October 1972), pp. 220–221.
30. 'Ariadne', *New Scientist* (13 April 1972), p. 104.
31. Williamson, D. T. N., 'The Anachronistic Factory', Paper to Royal Society, London (March 1972).
32. Burbidge, J. L., *Chartered Mechanical Engineer* (February 1973), pp. 74–76.
33. Aronson, T. F., 'Roots of justification for automation', *Automation* (November 1971), pp. 38–42.
34. Bennett, K. W., and Williams, D. N., 'Watch the robot plug himself in', *Iron Age* (April 16 1970).
35. *The Times*, London (30 March 1973).
37. Prenting, T. O., and Kilbridge, M. D., 'Assembly: last frontier of automation', *Management Review*, **55**, 2 (1965), pp. 4–19.
38. Iredale, R., 'Automated assembly—the state of the art', *Production Engineer* (July 1969), pp. 291–305.
39. Gerity, G., 'Automatic assembly reaches ahead', *Tooling and Production*, **33**, 6 (1967) pp. 67–71.
40. Given, K., 'The unmaking of the Vega', *Motor Trend* (May 1972), pp. 49–52, 128.
41. 'Building the Chevrolet Vega', *Automobile Engineer* (November 1970), pp. 456–461.
42. Rothschild, E., 'G.M. in more trouble', *New York Review* (23 March 1972).

43. Williams, D. N., 'Lordstown plant: G.M.'s new mark of excellence', *Iron Age* (11 March 1971).

44. Camp, C. B., 'Paradise lost', *Wall Street Journal* (31 January 1972).

45. Salpukas, A., 'G.M.'s Vega plant closed by strike', *New York Times* (7 March 1972).

46. 'G.M.'s design for easier assembly—the Vega 2000', *Assembly Engineering* (October 1970), pp. 26–29.

47. Rezler, J., *Automation and Industrial Labor*, Random House, 1969.

48. Buckingham, W., *Automation: Its Impact on Business and People*, Harper Row, 1961.

49. Daniel, W. W., 'Automation and the quality of work', *New Society* (29 May 1969).

Part 2

Dimensions of Development in Mass Production

The nature and effects of automation and mechanization in the context of mass-production assembly work, and the nature, means and objective of job restructuring.

CHAPTER 4

Automation and Mechanization of Assembly Work

In this chapter we shall examine the nature, pattern of development, characteristics and effects of mechanization and automation in mass-production assembly work. The general nature and effects of mechanization and automation will be examined together with the influence of certain technological factors. Certain case examples will be provided and particular attention will be given to the role of the worker in such systems and the nature of manual jobs.

THE NATURE OF MECHANIZATION AND AUTOMATION

Bright,[1] attempting to overcome some of the problems of definition, suggests that in general usage automation suggests something significantly more automated than previously existed. Thus automation is seen as a developmental process rather than a state; hence 'automation' in one industry may contrast both in level of development and characteristics with automation elsewhere. Bright identifies five 'fronts' on which this development takes place, i.e. materials, production processes, factory layout, product design and automatic machines, whilst other authors have taken the view that the elimination of direct manual involvement in control procedures is the principal feature of automation.[2,3]

Many authors[2,3,4] recognize mechanization as an aspect of automation, the latter being seen also as a function of the integration of processes or operations and the use of control systems which are largely independent of manual involvement. Jaffe and Froomkin[5] for example reserve the term automation for that type of process in which the automatic-feedback principle is used, whilst referring to mechanization as any technological change which increases output per worker.

It seems reasonable therefore to view both mechanization and automation as trends rather than states, in both cases the trends being associated with the replacement of human activities by the activities of inanimate objects. Mechanization, for our purposes, is therefore seen as an aspect of, or component part of, automation, mechanization being concerned with activities whilst automation of such activities implies the use of control procedures which are also largely independent of human involvement. These two terms will be used in this manner throughout the remainder of this chapter, i.e. automation will be taken to subsume mechanization.

Faunce[6] identifies four basic components of production systems in order to aid examination of the nature and effects of automation, i.e.

(1) Power technology—sources of energy used.
(2) Processing technology—tools and techniques used in actual operations performed on materials.
(3) Materials handling technology—transfer of materials between processes.
(4) Control—regulation of quality and quantity output.

He suggests that the substitution of inanimate for human performance of function occurs usually in the order (1) processes, (2) handling and (3) control.

Bright[1] has identified seventeen levels of mechanization and using these levels has plotted mechanization profiles for several production processes. This hierarchy of levels seems to be limited by its unidimensionality, the hierarchy necessarily including elements of processing, handling, etc. For descriptive purposes it is surely beneficial to identify hierarchical levels within each of the three components of the system, i.e. processing, handling and control. Such a principle has been used in the development of the hierarchies shown in Exhibit 4.1.

PROCESS (i.e. Assembly, Forming, Cutting, Measurement, etc.)

Level 1. Hand and handtool
     2. Powered handtool
     3. Machine—hand controlled
     4. Machine—automatic cycle—hand activated
     5. Machine—automatic

HANDLING (i.e. Transport, Load/Unload, Location, Storage)

Level 1. Hand and handtool
     2. Powered handtool
     3. Machine—hand controlled
     4. Machine—automatic cycle—hand activated
     5. Machine—automatic

CONTROL (i.e. (a) Activation, (b) Monitoring, (c) Regulation, and (d) Rectification and Maintenance of Processes and Handling)

Level 1. Manual (a), (b), (c) and (d)
     2. Product activated or timed with manual (b), (c) and (d)
     3. Automatic (a), (b), Manual (c), (d)
     4. Automatic (a), (b), (c), Manual (d)
     5. Fully automatic

Exhibit 4.1 Dimensions of automation

THE INFLUENCE OF TECHNOLOGY

If automation is developmental by nature then the comparative advantage and ease of provision of manual, mechanized and automated operation for each of the system components (processing, handling and control) will largely decide the nature of this development. The advantages of, and ease of provision of, mechanization and automation is to some degree influenced by factors such as the nature of the process and the materials being processed.

Materials and transformation processes may, for our purposes, be classified according to material characteristics, i.e. solid/fluid, and nature of material flow, i.e. intermittent/continuous. A further distinction based on energy requirements of transformation permits the distinction between certain solid-item transformation processes.

The transformation process in assembly involves the placing together of two or more parts in their correct relative positions. The energy requirement is determined by the physical size and weight of the components, a factor which also influences (although not necessarily determines) the nature of the material flow. In general, energy requirements in assembly are low whilst energy requirements in metal cutting are higher, a difference which can be seen as a major influence on the development of automation in these two technologies.

It seems clear that the developments within a particular technology, as they affect each system component, are relevant in the analysis of the nature of mechanization and automation, the organization of work and the nature of the work environment.

For our purposes an important area of study in the assessment of the effects of automation is the individual's job. It seems reasonable to accept the position that tasks are 'embedded' in the technology and consequently technology or process factors are again seen as constraining influences on the nature of jobs and job design. We cannot therefore take a deterministic view on the nature of the influence of technology on tasks, rather we must accept the position advanced by Wedderburn and Crompton[7] who suggest that jobs are influenced by the constraints placed upon them, one of which is technology.

Comparison of metal cutting and assembly

In machining, as in assembly, work flow is intermittent and the material processed is solid. The energy requirement is much higher in metal cutting and thus the requirement of large quantities of accurate parts quickly give rise to process mechanization. During the early part of the eighteenth century the transformation process was developed to stage three (Exhibit 4.1), i.e. machine–hand controlled, and development to stage four followed quickly. The mechanization of handling was developed in the mass-production metal working where relative inflexibility was acceptable. The first transfer-machine was developed for woodworking in 1908, whilst Archdale built a metal-cutting transfer machine as early as 1922–1923 (Wild[8]). Thus materials handling and work-in-progress costs were reduced by mechanization.

The energy requirement of the assembly process is relatively low, whilst in comparison the higher energy required to move large quantities of subassemblies and assemblies made material movement the first area for mechanization. The transformation process in assembly remains largely unmechanized although in some industries mechanized assembly has been applied for many years. Such industries (e.g. electric-lamp manufacture) are characterized by high, stable demands, thus special-purpose equipment is economically viable.

In industries such as automobile-subassembly manufacture, where demand is high and stable, general-purpose assembly machines have generally replaced manual flow lines. Consequently the mechanization of the transformation process has again followed the mechanization of the handling process. Many such applications integrate manual and mechanized stations into one system using a non-synchronous or free-transfer system (Boothroyd and Redford[9]). Assembly machines usually incorporate 'sensing' devices and 'memory' systems for in-process inspection purposes and consequently control is developed to level 3.

Technological development in the small- and medium-batch assembly area is limited by the need for flexible control and peripheral devices able to deal with components of varying size and shape (Heginbotham[10]). The use of easily programmable robots used for example on positioning and spot-welding tasks is increasing and machines controlled by electronic logic have been developed (Heginbotham[10]). However, in general, the variability associated with such production ensures a low level of mechanization is employed although efforts to improve the effectiveness of power tools and other mechanical aids are clearly greatly important.

Case examples in assembly automation

In order to help identify the factors affecting development and the nature of the development of automation assembly, three case examples will be examined.

Case 1: printed-circuit-board assembly

This involves the positioning, insertion and fixing of various electronic components to circuit boards. The operations, originally performed manually, are:

Component forming.
Insertion.
Cropping (of wires to length after insertion).
Soldering.

Only insertion and soldering are specific assembly tasks, the others being essential component-preparation tasks. Mechanization was first applied to these preparation tasks; thus components are formed and cropped mechanically and all soldered contacts made simultaneously and mechanically (flow soldering). Assembly tables of various configuration were introduced with various component-holding devices to facilitate manual component selection and positioning for insertion. Further developments are appropriate for batch production (i.e. mechanization capable of being 'programmed' to take account of variability) or for very long runs. The latter case covers the introduction of automatic transfer-line-type insertion machines using ready-formed and cropped components fed into such machines on 'bandoliers'. The former area is covered by several levels of sophistication, i.e.

(a) Hand-assisted, mechanized component insertion (e.g. pantograph).
(b) Numerically controlled insertion machine.
(c) Computer controlled insertion machine.

Thus development appears to have taken place as follows;

(1) Mechanization of preparation processes (forming, cropping, soldering).
(2) Mechanization of handling for insertion.
(3) Mechanization of insertion with an associated integration of preparation and insertion (if only by use of bandoliers).

Case 2: vehicle-body assembly (Volkswagen, Stalhuth[11])

Vehicle-body assembly in this case may be divided into front section, rear section, roof assembly and final assembly. Various processing operations (e.g. spot welding, press work) are associated with these assembly 'operations'. Initially, processing was undertaken largely manually, parts being manually clamped into jigs prior to manual welding. Handling, however, was well developed with fairly complex powered conveyor systems in general use. Between 1955–1958 rotary-indexing machines were introduced for front section, rear section and roof assembly. These machines were semi-automatic, incorporating mechanized spot-welding stations with manual loading. Completed assemblies were unclamped and lifted automatically from the jigs in the last station and each machine producing 240 assemblies per hour was operated with a five-man crew. Thus by 1958 only the three main sections—front half, rear half and roof—had to be clamped into the body jigs by hand. Attention was consequently focused on final assembly, leading eventually to the introduction (in 1963) of a multi-station transfer machine in which the three main subassemblies are automatically loaded, fitted together and welded into a unit. This system integrates presses, automatic spot-welding, automatic and manual loading. Final welding of the undersection, where weldments are required in areas too inaccessible to be reached by the automatically operated equipment of the main body jig, is performed manually. With a production rate of 200 body shells per hour, the system released 440 men and 88,000 square feet of floor-space.

It will be noticed that processes (press-work and handling) were initially at a high degree of mechanization. Developments of mechanized spot-welding on rotary-indexing machines followed. Finally, assembly was mechanized to provide a highly mechanized, integrated manufacturing system in which mechanized and manual operations are integrated and for which handling has been developed to a sophisticated level of mechanization.

Case 3: electric-lamp manufacture (Bright[1])

The manufacture of electric lamps involves the use of widely different materials (metals, glass, gas), some extremely fragile components, a high degree of precision, the need to test each complete fragile end-product. Initially manufacture was carried out on a batch-production basis. The operations required were set out on a sequential basis but no provision was made for sequential movement.

During the period 1920–1925 rotary-indexing machines were introduced to reduce work-in-process inventory and improve product quality. Later (1925–1936) the material movement between machines was mechanized and finally the more difficult mounting operation was mechanized. Examining these developments, Bright noted the following important trends:

(1) Mechanization of manually performed operations.
(2) Integration of machines with automatic work-feeding, work-removal and material-handling devices.
(3) Combination of different operations on a single machine base.
(4) Changes in product design to permit mechanical manipulation.

These three cases demonstrate certain important differences in assembly automation. For example, in the case of printed-circuit-board and electric-lamp assembly, mechanization of processes was introduced to situations of low-handling mechanization, whilst in the automobile-body-assembly case the mechanization of handling had already reached a relatively high level of sophistication before the introduction of the rotary-indexing machines. Despite such differences certain trends emerge and are indicated on Exhibit 4.2, in

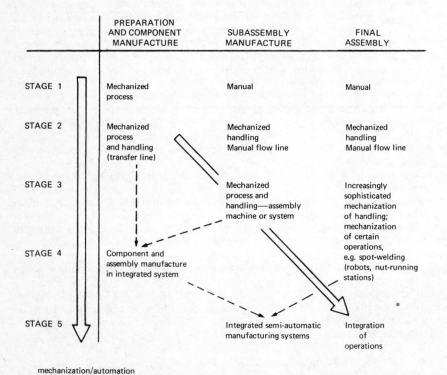

Exhibit 4.2 General development of assembly automation

which we have split discrete-item manufacture into three stages:

Preparation and component manufacture.
Subassembly manufacture.
Final assembly.

It seems reasonable to suggest that mechanization has been introduced initially at the preparation stage, then subassembly and finally at final assembly (i.e. from low complexity to high complexity of manufacture). This exhibit records developments that have taken place or that are clearly 'heralded' by manufacturers and researchers in the area of production technology. We do not suggest that all mass-production assembly will develop in exactly the same way but simply that the exhibit should allow us to 'map' any developments that have taken place in assembly automation.

In assembly and metal-cutting, development of mechanization in processing or handling appears to have been largely dependent on process factors. It is clear that mechanization of processing and handling has brought about the integration of operations and a tendency toward the development of semi-automated systems (e.g. the body-assembly and electric-lamp assembly cases).

THE EFFECTS OF AUTOMATION

The effects of both increasing mechanization, and trends towards process automation, on the skills and behaviour required of workers, supervisory patterns, etc. have been studied extensively. Such research has relied on comparative studies of production systems commonly held to represent different degrees of mechanization or automation, as well as studies conducted within one industry.

Automated systems

Studies of the effects of automation have typically involved examination of work undertaken in process industries. Mann,[12] reviewing earlier studies[4] in power plants, and making comparisons with studies of automation in steel mills (Walker[13]), observed that automation in such situations tended to give rise to wide basic changes in the content and structure of jobs, working conditions, career patterns, security, pay and the prestigiousness of the work performed. With increasing automation jobs were found to provide more demanding, varied, interesting and challenging work for many workers, although in some cases such changes were suspected to be of a temporary nature—a result of a 'start-up' situation. Technical know-how was found to be relatively of more importance. Daniel,[14] also studying process plants, confirmed many of these observations, whilst Hardin and Byars[15] suggest that workers may expect increased job content resulting from automation, together with increased demands on skills, knowledge and training. Similar views are expressed by other researchers (e.g. Davis,[16] Drucker[17] and Emery and Marek[18]) who in general report or expect greater complexity and responsibility, and therefore intrinsic rewards, to be associated with work in automated systems, but often at the expense of increased worker inactivity.

Many of the authors dealing with the effects of process automation deal also with social interaction. Mann[12] identifies the greater distance between workers as resulting in reduced interaction. However, he excludes the case of full process automation where grouping of controls gives rise to grouping of workers. Foster[2] supports this general view in stating that advances in automation increase the ratio of working space to people and therefore reduce social relationships. The relationship of operators and their supervisors is also considered to be affected by automation, the general view being one of increased contact and improved worker/supervisor relations. An increased separation of workers from both processes and their products is noted by several researchers (e.g. Emery and Marek[18]). Increased training needs have been associated with the wider responsibilities of automated jobs whilst emphasis on vigilance and monitoring duties, the importance of minimizing process disruption, the consequences of breakdowns and the comparative inactivity of workers are considered on occasions to lead to increased stress.

Mechanization

Although comparatively little work has been undertaken, in general researchers (Davis,[16] Faunce,[19] Buckingham,[3] Bright[1]) conclude that the introduction of mechanized systems gives rise to

(1) The increasing isolation of workers and hence a reduction in social interaction.
(2) A reduction in the amount of physical effort required, largely due to reduced handling requirements.
(3) A loss of worker control of work-pace and worker independence from the machine cycle.
(4) Improved working conditions and increased safety, and
(5) Increased use of shift working.

The degree to which workers' jobs are affected is one topic on which authors have been at pains to distinguish between the effects of mechanization and automation. Bruyns,[20] for example, suggests that the principal effect of mechanization is to restrict workers' actions, whilst automation takes over these actions. Davis[16] lends support to this view in indicating that workers who are 'in-line' with the production system may have work tasks which have the repetitiveness and tensions of similar non-mechanized work. This view is also supported by Buckingham[3] and Faunce[6] who suggest that in many cases mechanization adversely affects jobs by increasing the division of labour, rendering certain skills obsolete, removing control of the work-pace and increasing the 'distance' between worker and product. It seems to be generally accepted that in both mechanization beyond a certain stage and automation the emphasis on inspection, monitoring and control tasks increase whilst the amount of direct production activity decreases. For example the degree of mechanization involved in transfer machining (Faunce[19]) is considered to provide jobs with increased responsibility requiring more alertness and perhaps therefore greater fatigue. Higher levels of mechanization appear to necessitate the use of greater skills, wider knowledge and the performance of supervisory duties and thus may differ little from the effects of automation. However to achieve this situation the various stages of partial mechanization, characterized by the retention of some manual work in either processing or handling appears to necessitate a somewhat different role for the worker. Bruyns[20] suggests that the principal characteristic of the change from mechanization to automation concerns process-control activities. Thus he considers that many so called mechanized processes required the worker to exercise control by

(1) Supervise a self-activating production system.
(2) Service, maintain and eliminate disturbances.
(3) Ensure operation within fixed limits.

He considers that such characteristics imply that the part of the process controlled by the worker is larger than that part previously operated by the worker, that there is a shift towards evaluation and working with abstract information, that communication takes up more time and that the gap between the worker and the process is widened. Such characteristics require the worker to have a knowledge of the whole process, powers of imagination and combination, and the ability to bear stress caused by the intricacy of the process, the lack of direct control, sudden disturbances and the isolation of the job. Such changes clearly give rise to an enlargement and an enrichment of work although offset perhaps to some degree by relative inactivity and social isolation.

Work roles

The information presented above is summarized in Exhibit 4.3, which is intended to provide a *general* picture only of the relationship between certain work tasks and the levels of automation. The classification adopted previously is employed in order to provide examples of production systems at various levels of mechanization/automation.

Whilst the manner and characteristics of the development of automation are affected by the factors identified in the previous section, researchers in the area appear to be in general agreement that at higher levels of automation work becomes more varied and demands greater use of skills and knowledge, offset to some degree by physical inactivity coupled with the need for vigilance which could give rise to a stressful situation. Equally, it is clear that what has generally been referred to as mechanization, because of continued dependence on manual intervention in control, offers few of these job characteristics. The cases presented previously suggest that in most cases advanced assembly systems are at best examples of advanced mechanization rather than automation, and hence it would appear that in general the workers associated with such systems may not benefit from some of the job improvements commonly associated with continuous-process automation.

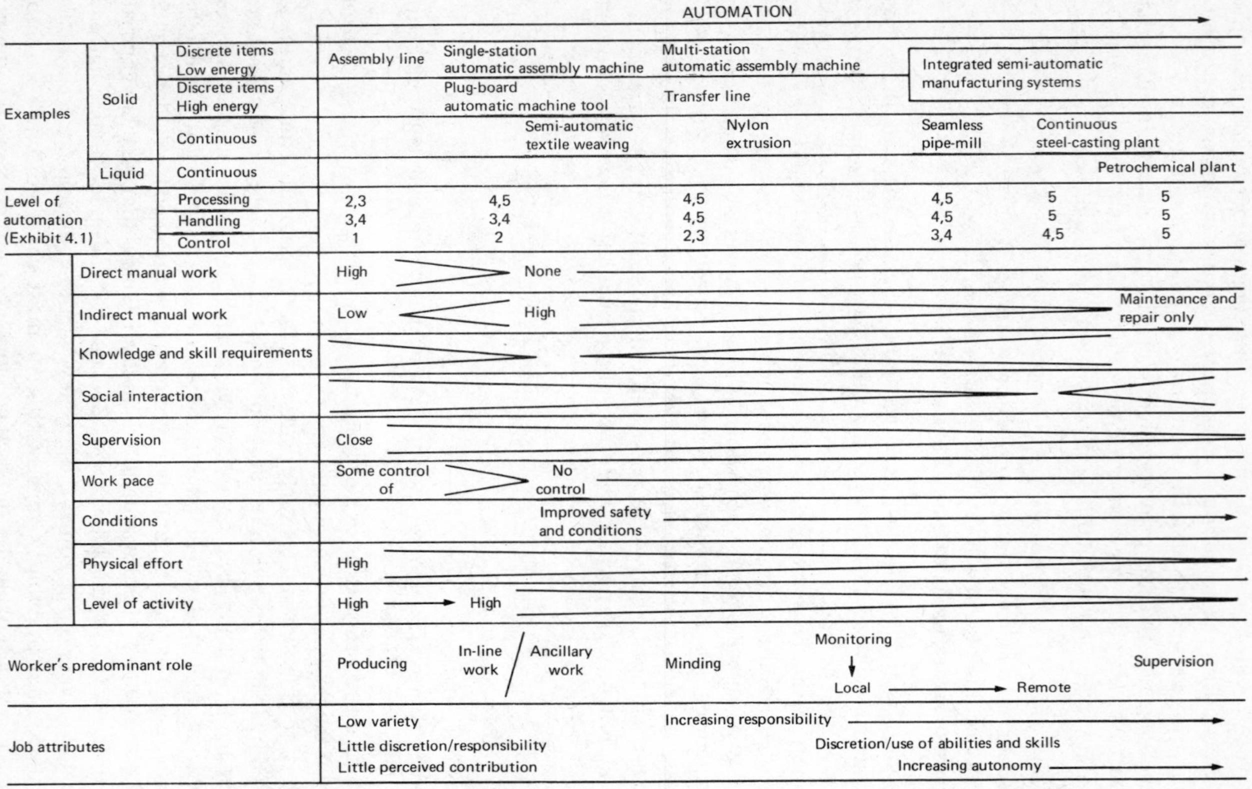

AUTOMATION →

		Assembly line	Single-station automatic assembly machine	Multi-station automatic assembly machine		Integrated semi-automatic manufacturing systems	
			Plug-board automatic machine tool	Transfer line			
Examples			Semi-automatic textile weaving	Nylon extrusion	Seamless pipe-mill	Continuous steel-casting plant	Petrochemical plant
Solid	Discrete items Low energy						
	Discrete items High energy						
	Continuous						
Liquid	Continuous						
Level of automation (Exhibit 4.1)	Processing	2,3	4,5	4,5	4,5	5	5
	Handling	3,4	3,4	4,5	4,5	5	5
	Control	1	2	2,3	3,4	4,5	5

Direct manual work: High ————→ None

Indirect manual work: Low ————→ High (Maintenance and repair only)

Knowledge and skill requirements

Social interaction

Supervision: Close

Work pace: Some control of ————→ No control

Conditions: ————→ Improved safety and conditions

Physical effort: High ————→

Level of activity: High ————→ High

Worker's predominant role: Producing | In-line work / Ancillary work | Minding | Monitoring | Supervision

Local ————→ Remote

Job attributes:
Low variety ————→ Increasing responsibility
Little discretion/responsibility ————→ Discretion/use of abilities and skills
Little perceived contribution ————→ Increasing autonomy

Exhibit 4.3 Nature of work at various levels of automation

ASSEMBLY AUTOMATION AND THE WORKERS

The previous discussion suggests that discrete-item assembly automation has yet to progress substantially beyond the stage of process and handling automation, hence even the more advanced system still bears little resemblance to the classic areas of process automation. Assembly systems therefore continue to depend upon manual intervention ranging from manual production work in, for example, assembly-line systems to manual override control on multi-station 'automatic' assembly machines. Since manual work, whether direct or indirect, appears likely to continue to be an important characteristic of assembly systems we will in concluding this chapter examine the nature of such work under the influence of increasing automation.

The work roles associated with the main levels of assembly mechanization identified in Exhibit 4.3 are summarized below and in Exhibit 4.4.

Level 1: the manual-flow-line worker

The role of a worker on a manual flow line is that of *'producing'*. His job has elements of repetitiveness and pacing influenced by the line cycle time, method of operation and use of buffer stocks. Often work will be performed with the assistance of power tools and interstation handling will be mechanized. The job often involves a relatively high physical content, and functional interaction may be limited. Social interaction, possibly only with immediately adjacent workers, will be influenced by factors such as degree of pacing and line length.

Level 2: in-line and ancilliary work

In some cases workers perform assembly or ancillary tasks at a work-station or at a single-station machine, typically where mechanization of the operation is unjustified on economic grounds. Such a worker may be considered to be 'in-line' with the system and is likely to be subjected to the pacing, work pressure and repetitiveness characteristics noted by research workers investigating manual flow lines. Furthermore, such characteristics may be accentuated by the very short cycle times typical of such machines. Additionally, conditions of social isolation, reduced social interaction, reduced physical effort, loss of worker control of work-pace and worker independence from the machine cycle may also prevail.

Level 3: machine minding

Work at this level is predominantly that of machine minding which normally requires:

(a) Loading parts or assemblies into the machine.
(b) Observing the progress of assemblies at the various stations of the machine by making direct visual checks or by watching a monitoring panel.

Level	Description	Tasks
1	Flow-line assembly worker. Manual processing (power tools). Mechanized handling	Repetitive, short-cycle (1–3 min.). Pacing. Relatively high physical content
2	Worker 'in-line' with assembly machine or system. Manual processing. Mechanized handling	Repetitive, very short cycle (5–10 secs). High degree of pacing. Low physical content
3	Machine-minder. 'Operates' assembly machine. Load/unload, de-jam workheads, simple maintenance tasks. Intermediate mechanization of processing, handling. Low mechanization of control	Mix of repetitive work (e.g. load/unload) and non-repetitive response work. Some emphasis on knowledge of operation of pneumatic/hydraulic/electrical equipment. Low physical content. Increased attention span
4	Machine-monitor. Integrated manufacturing system. Relatively more sophisticated level of mechanization of control. Integration of all or part of component production, subassembly manufacture, final assembly	Mix of repetitive (e.g. load/unload, making good faulty assemblies) and non-repetitive monitoring associated with greater sophistication of control. Membership of team responsible for operation of system. Emphasis on broader skill profiles. Relatively large attention span. Increased responsibility. Low physical content

Exhibit 4.4 Work roles associated with the main levels of mechanized assembly

(c) 'De-jamming' workheads.

(d) 'Making-good' faulty assembly.

The need for machine minding stems from the combination of high process and handling with low control automation. The former requires the use of greater skills and wider knowledge to ensure control of the system through

(1) Supervision of a self-activating system.

(2) Service, maintenance and elimination of disturbances.

(3) Ensuring operation within fixed limits (Bruyns[20]).

Such activities will in general be carried out with limited social interaction, perhaps comparatively long periods of inactivity and hence possible increased tension.

Level 4: machine monitoring

The use of manufacturing systems incorporating high levels of mechanization and integration together with the associated high degree of sophistication in control mechanization will necessitate close machine monitoring. This job, probably largely carried out remote from the manufacturing system, is likely to provide a mixture of fairly repetitive work interspersed with non-repetitive monitoring tasks. This may result in comparatively long periods of relatively low-level activity, low physical activity, whilst requiring relatively large attention span and providing close interaction with other members of a mixed-skilled team. Such interaction will be functionally based, supervision may be low and potentially a high degree of social cohesion may be possible.

References

1. Bright, J., *Automation and Management*, Division of Research, Harvard Business School, Boston, 1958.

2. Foster, D., *Automation in Practice*, McGraw-Hill, 1968.

3. Buckingham, W., *Automation*, Harper and Row, 1961.

4. Mann, F., and Hoffman, L. R., *Automation and the Worker*, Henry Holt, New York, 1960.

5. Jaffe, A. J., and Froomkin, J., *Technology and Jobs*, Praeger, 1968.

6. Faunce, W. A., 'Automation and the division of labour', *Social Problems*, **13** (Fall 1965), pp. 149–160.

7. Wedderburn, D., and Crompton, R., *Workers' Attitudes and Technology*, Cambridge University Press, 1972.

8. Wild, R., *Mass-Production Management*, Wiley, London, 1972.

9. Boothroyd, G., and Redford, A. H., *Mechanised Assembly*, McGraw-Hill, London, 1968.

10. Heginbotham, W. B., 'Automatic assembly tomorrow', *The Production Engineer*, **49**, 7 (1970), pp. 282–288.

44

11. Stalhuth, W. E., 'Evolutionary role of automation in assembling body shells', *Automation*, **11**, 2 (1964), pp. 69–75.

12. Mann, F. C., *Psychological and Organisational Impacts in Automation and Technological Change*, The American Assembly Spectrum Books, 1962.

13. Walker, C. R., *Toward the Automatic Factory*, Yale University Press, 1957.

14. Daniel, W. W., 'Automation and the quality of work', *New Society*, **13** (1969).

15. Hardin, W. G., and Byars, L. L., 'Human relations and automation', *S.A.M. Advanced Management Journal* (July 1970), pp. 43–49.

16. Davis, L. E., 'The effects of automation on job design', *Industrial Relations*, **2**, 1 (1962), pp. 53–73.

17. Drucker, P. F., *The Practice of Management*, Harper.

18. Emery, F. E., and Marek, J., 'Some socio-technical aspects of automation', *Human Relations*, **15** (1962).

19. Faunce, W. A., 'Automation and the automobile worker', *Social Problems*, **6** (1958), pp. 68–78.

20. Bruyns, R. A. C., 'Work and work motivation in an automated industrial production process', *Management International Review*, **19**, 4–5 (1970), pp. 49–64.

CHAPTER 5

Job Restructuring

In this chapter we shall examine the concepts, nature and practice of what will be referred to, for convenience, as *job restructuring*, i.e. that development which concentrates upon the redesign of previously rationalized, repetitive and constrained jobs. We shall tend to deal with the subject from the point of view of changing an existing situation, i.e. *re*structuring and *re*design rather than structuring or design. Such an orientation is convenient, since most of the information that we shall draw upon relates to such changes. However, most, if not all, of what will be considered applies equally to the original design as well as the change situation, both being of importance in connection with the development of mass-production systems.

A survey approach will be employed in order to identify the basic techniques of job restructuring as well as the reasons for, effects of and scope for, such techniques. In the last section a similar approach will be employed in order to develop views on the limitations on job restructuring, and finally to develop a detailed checklist or model to summarize both the means by which jobs might be restructured and the objectives of such changes.

TECHNIQUES, REASONS AND SCOPE OF CHANGES

The objectives of this section are pursued through an examination of the published accounts of restructuring exercises and experiments undertaken in industry. Other published information is employed where relevant in interpreting this survey, which is confined to blue-collar, discrete-item manufacturing jobs.

Survey of exercises

Brief details of the exercises and experiments reported in recent literature are given in Appendix A together with source references. These ninety-six exercises are categorized according to the principal type of change which had been introduced. Frequently exercises involved the introduction of multiple changes and in such cases classification is based on what appeared to be the major change. The following principal types of change are identified.

(1) *Rearrangement/replacement of assembly-line work*

In these thirty-four experiments manual-assembly-line production systems were substantially modified and often replaced by other systems of working. In all cases the revised job was free from mechanical pacing, and in many cases workers were also made responsible for the inspection of their own work. All examples in this category relate to the production of discrete items and in general the work involved in such processes is unskilled, often being undertaken by women workers. The principal characteristic of these exercises is the increase in the size of the tasks undertaken by assemblers together with a relaxation of the constraints previously imposed by the flow-line discipline.

(2) *Workers given additional responsibility*

In these thirteen cases jobs have been redesigned with the primary intention of increasing the responsibilities of the individual worker or the work group. Such increased responsibilities include inspection of own work, completion of orders and involvement in some other specific aspects of the job. In many cases the purpose of these experiments was the improvement of output quality through the greater involvement of the individual with his job. Some of the exercises relate to the jobs of flow-line workers, but in general these efforts to increase responsibilities have mainly concerned workers operating independently.

(3) *Rotation of jobs*

In these eight exercises the workers were given the opportunity to perform more jobs than the one on which they had originally been employed. Most exercises related to repetitive flow-line-type jobs and in many cases additional training was given to workers prior to the introduction of the change. In most cases such job rotation was seen as a means of increasing the flexibility of the labour force and of the production system, although in certain cases the principal motives for the change related to increased challenge and variety in the work.

(4) *Responsibility for additional and different types of work*

In these five exercises the jobs of workers were modified in order to add to the number of different tasks performed and thus to utilize further worker skills. In several cases the addition of tasks involved the allocation of responsibilities for indirect work, such as maintenance and tool setting.

(5) *Control of work speed*

In several of the exercises included in the above categories workers assumed greater control of their own work speed. This change normally results from the rearrangement of assembly-line work ((1) above); however, additionally one exercise is reported in which workers on an assembly line were allowed to decide amongst themselves the speed of the line. Although this did not remove the pacing effect it did give workers some overall control of their work pace.

(6) *Self-organization*

In fourteen of these experiments workers were released from external control and influence over comparatively long periods of time. Such workers were normally engaged on the assembly of a complete unit or subunit, the manning of certain equipment or were responsible for a complete work area. They were usually made responsible for additional indirect tasks such as quality inspection, and to some degree also for work organization and planning. Such workers were usually able to perform most if not all tasks undertaken in a particular work area and in five cases either individuals or the groups of workers were involved in problem-solving activities. Such changes were often preceded by training sessions.

Types of change

The changes cited above cover two basic facets of job restructuring, i.e. the *enlargement* of jobs through the addition of one or more related tasks, and job *enrichment* involving the increase in the motivational content of jobs through, for example, the addition of different types of task, or the provision of increased worker involvement and participation. This useful distinction between *horizontal* (enlargement) and *vertical* (enrichment) job changes is

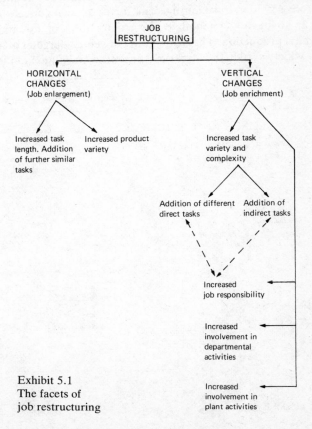

Exhibit 5.1
The facets of
job restructuring

48

illustrated in Exhibit 5.1. On reflection we see that the work categorized under (1) above constitutes both enlargement and enrichment whilst categories (2), (4) and (5) constitute some form of vertical change, i.e. job enrichment, in most cases confined to an increase in the individual's involvement in his job through increased job responsibilities. We can also see that almost a quarter of the exercises reviewed, i.e. categories (3) and (6) conform to neither of these facets, the reason being that the structure shown in Exhibit 5.1 applies only to job changes, whilst neither job rotation (3) nor 'self-organization' (6) *depend* upon the manipulation of individual jobs. Indeed it can be argued that neither job rotation nor self-organization is primarily concerned with job restructuring and that they represent a different form of change. Such distinction is appropriate since although this further form of change may give rise to, or incorporate, some degree of job restructuring, it would appear to have a somewhat different orientation, being concerned primarily with the reorganization of work, and of workers.

Thus it would appear that the title employed for this chapter in fact covers only one part of the relevant subject matter as there are *two* related aspects to the dimension of change with which we are dealing—job restructuring, where the emphasis is upon the redesign/design of individual jobs, and work organization, which, we shall hypothesize, gives rise to, facilitates or forms a prerequisite for job restructuring. This two-part categorization—Exhibit 5.2—appears to be fundamental to the examination of this dimension of change in mass production. The distinction between enlargement and enrichment terms widely used in the literature) permits examination of the degree to

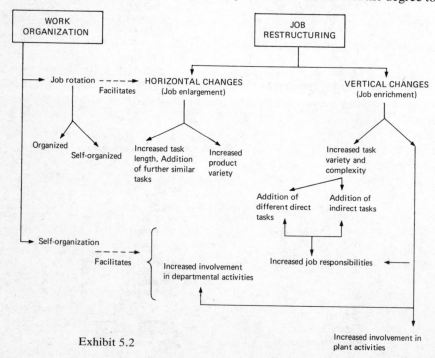

Exhibit 5.2

which the changes employed are likely to increase the motivational content of jobs. It is argued that the opportunity for the satisfaction of higher order needs[1] is provided through job enrichment but not through the simple enlargement or extension of the existing content of jobs.[2] The difference between job restructuring and work organizational changes help highlight the fact that although the objective of many changes is the modification of the tasks undertaken by workers, such changes are often dependent upon, or perhaps only brought about by, appropriate organizational change. Thus job rotation—an organizational change—may provide for job enlargement, and some degree of worker self-organization may give rise to, or be a necessary prerequisite for, certain types of job enrichment.

Reasons for, and the results of, changes reviewed

Of the thirty-two reasons given for the changes reviewed above, seven were of an economic nature, including the need for reducing costs or increasing productivity, twelve were related to personnel problems, including the need to reduce labour turnover and absenteeism or improve working relations, whilst five dealt with the quality of output. A similar pattern of results was obtained by Reif and Schoderbek,[3] who identified the following reasons:

To reduce costs	21
To 'enrich' the job for the worker	15
To decrease specialization	14
To improve quality	13
To reduce monotony	11
Others	18
	—
Total	92
	—

Similarly, Wilkinson,[4] researching into Western European experiments in motivation, listed reasons for exercises as follows:

To improve output	11
To eliminate social problems	11
To reduce turnover and absence	6
Commercial factors	4
Concern about the role of first-line supervisors	8

Wilkinson also reported that four out of the seven most successful applications in his study 'were instigated not as experiments in motivation, but as fairly conventional productivity improvement exercises which attempted to take into account not only the technical aspects of the business but also the most recent thinking on how the human resource should be handled'. Reif and Schoderbek also found that firms usually had multiple objectives when making such changes, the desire to make work more interesting and challenging often being secondary.

A number of the experiments listed in Appendix A were introduced as part of a productivity deal aimed at removing trade demarcations and increasing the flexibility of the work force, thus reducing the effects of absenteeism. Certainly, the experiments appear to have been aimed at the solution of long-term problems which have failed to be solved by adjustment of these factors described by Herzberg as 'hygiene factors',[2]

The benefits claimed for the restructuring exercise listed in Appendix A included:

(1) Greater productivity.
(2) Improved quality.
(3) Fewer grievances.
(4) Improved worker attitudes.
(5) Better absenteeism and turnover records.
(6) Lower costs.

In some cases, these benefits were quantified, but only in a few cases was there evidence of systematic study of the change and its effects.

Few abandoned or failed experiments have been reported. Exceptions to this are an experiment at Philips where job rotation of assembly-line workers failed due to inadequate preparation and the adoption of the wrong approach,[5] and the experiment at Hovey and Beard where workers were given control over the belt speed and achieved output and earnings above other similar groups, thus causing unrest amongst other employees.[6]

Some problems arising out of the changes are reported. Resistance to the changes was often manifest amongst supervisors, unless they had been involved from an early stage and were assured of their own security. Changes to the payment system were not always made following increases in the responsibilities of workers. Philips reported that the changes to the job design were often not popular with younger workers, who often preferred to do jobs which demanded the least of their attention. The need for good preparation was stressed in a number of cases.

The scope for change

Anderson[7] suggests that the change strategy that might be adopted depends to some considerable extent upon the technology of the industry and obstacles arising as a result of that technology. Bearing this in mind we will examine the scope for change in the context of three basic types of job, each associated with a different type of production system, i.e.

(1) Assembly-line work.
(2) Machine operating.
(3) Machine supervising.

(1) *Assembly-line work*

All assembly lines fall somewhere on the continuum between the heavy-assembly technology where, as well as heavy capital investment per worker, there is usually complex transfer equipment (e.g. the multi-station automobile line), and the low-capital-investment line with a simple transfer system.

In the heavy-assembly situation the job module is defined by the capital equipment that surrounds the work station. Changes to this equipment are very costly, and therefore are usually only made when the products are modified. Secondly, the job is physical, highly repetitive, and strictly machine-paced, thus it is often impossible to trace the source of poor workmanship, hence inspection of completed goods is essential as is the use of utility men to rectify incomplete or faulty work. Often the individual is unable to appreciate the contribution of his work to the completed product.

On the other hand, in the case of the low-capital-investment line, changes can be made relatively easily and are often required by changes in product design. Production is often in batches, work pacing is not normally rigid and the worker is better able to appreciate his contribution to the complete process. Comparatively less training is necessary for the worker to become proficient in all the tasks on the line.

It can be observed from reports of experiments which have taken place on heavy-assembly lines, that one of their aims appears to be the establishment of job modules that include a functional, indentifiable product. Secondly, an attempt has been made to provide speedy feedback of results by placing the inspection function nearby.

Attempts to form teams or groups have also been made in a number of cases. These groups were given responsibility for organizing the work including scheduling materials, rejecting faulty components, deciding 'who does what' and some participation in decision-making. In these industries, where any experimentation is very costly if unsuccessful, every method available is likely to be used to ensure a favourable outcome.

The majority of those experiments dealing with changes to assembly-line jobs involve products where the technology is simple and often, in such cases, the use of the assembly line might only be marginally beneficial.

The cost of experimenting with this type of line is not usually very high in relation to the possible benefits which can be obtained in increasing flexibility, reduced stocks, improved quality and increased efficiency as a result of the elimination of the balancing loss. Most of these experiments have taken the form of individual or small-group assembly with inspection of completed products or subassemblies.

(2) *Machine operating*

The machine-operator's job traditionally has been one of low-skill requirements, repetition and machine pacing in noisy conditions with little personal contact between workmates. The more skilled jobs of setting-up machinery, maintenance and inspection of the final product have generally been performed by operators with more technical training.

Job redesign for machine operators in most cases have involved the broadening of the job by the addition of work requiring new skills and, in some cases, involving the addition of new responsibilities such as inspection of the product.

This requires initial training but leads to higher job status and higher wages for the operators concerned.

(3) *Machine supervising*

This type of work is usually carried out by a number of operators each with his own specific trade or skill. Often the skills required for each job are similar but rigid demarcation restricts cooperation. Restructuring in most of these experiments was aimed at a rationalization of traditional work roles and a reduction in rigid demarcations. The new jobs often contained responsibility for quality, light maintenance, some decision-making and work scheduling.

LIMITATIONS ON RESTRUCTURING AND ORGANIZATIONAL CHANGES

In this section we shall look in more detail at the *limitations*, *constraints* and *difficulties* associated with changes. These are examined mainly by reference to the results of the reported exercises. Most exercises have primarily involved job restructuring and this emphasis is reflected in our discussion.

Comparatively few writers have dealt with the limitations on changes either in respect of the type of work to which changes might usefully be made, the methods of introducing such changes or the subjects most likely to respond favourably. Any type of job is thought to be suitable for enrichment by some authors who suggest that since the attitude of the subject to enriched work is unpredictable there is little justification for selecting personnel, although senior job holders are considered to be more suitable.[13] It has been stressed that before being placed in an enriched job the individual differences of the subjects must be considered.[7,14] In particular, older workers are possibly more reluctant to accept such changes,[15] a greater degree of success having been found among younger workers in some cases,[16] but not in the Philips Company.[11] It has been suggested[17] that job enrichment is inappropriate for those persons who 'seek to minimize effort expended on the job', a view which offers a possible explanation for the findings of the Philips organization with reference to younger workers. It is thought preferable that job enrichment be introduced to groups of coherent similar workers, i.e. single sex, similar skill levels, and also to jobs at high skill levels, and it is also suggested that where piecework bonus payments exist satisfactory job changes are difficult to implement.[16]

The problem of overcoming resistance to change has been discussed by several writers, most of whom conclude that early commitment to success by those concerned in the change can only be achieved by their early involvement.[18] It has been emphasized that before job restructuring is undertaken supervisors must be fully committed to the achievement of a successful outcome, and that sympathetic members of top management should be convinced first in order to ease the problems of implementation.[19,20] Thus, if jobs are to be restructured *all* the factors affected must first be carefully considered,[17] including the changed training needs.[3,7] It has been observed that attempts to change jobs may bring resistance to change,[3] and thus the chances of success may be increased if changes are introduced as part of a larger exercise which includes efficiency and productivity drives.[4] The early involvement of operators is advocated,[20,21] possibly at the stage when the jobs are being designed, and change in managerial style (to the more participative) is seen as a way of overcoming resistance,[22] some of which, however, may originate from Union involvement.[3]

Job changes at one level are seen by many writers to affect jobs at higher levels. It has been recommended that work be shifted down from managers,[20] hence managers must be willing to delegate more autonomy which may require the development of a new managerial style.[21] The supervisors' position and status may well be impoverished;[4,16] they should however have more time available for 'important work'[7,13] but some may not accept their loss of direct authoritative control.[20] The jobs of others may be affected by the restructuring of some jobs[20] and it is necessary to consider the change in status of all those involved.[17]

It has been proposed that 'brainstorming' meetings be held in order to develop ideas,[13] and that workshop sessions to solve problems selected by a team leader or key worker within the department may be employed.[7] It has been suggested that if workers are within autonomous groups there is more likelihood of the desired job characteristics being present than if they work individually, especially if there is job rotation within the group.[4] The benefits of job rotation, however, on occasions can be outweighted by the disruption of social relationships.[4]

Job enrichment is seen to be an investment of current expenditure in order to achieve long-term benefits.[20] This may lead, however, to demands for higher pay and conditions which are considered appropriate for jobs with higher responsibility.[16]

THE MEANS AND THE ENDS OF CHANGES

Our primary objective here is to attempt to develop a succinct yet comprehensive model or checklist which can be used in examining the case examples presented later. In particular we must look in more detail at the mechanics of restructuring and organizational changes, i.e. the 'means', and at the objectives, i.e. the 'ends', in such change.

For this purpose most major published views relating to means, objectives and effects were examined where such views were specifically concerned or were judged to be relevant to blue-collar jobs. In order to provide a common and practical framework, information was interpreted, and is presented in respect of

(1) *Job characteristics*, i.e. those features/aspects of jobs which might be manipulated or provided, and
(2) *Objectives or effects*, i.e. worker responses to changes or the effect of such changes.

Following the observations made earlier, such job characteristics are grouped as follows:

Job/Work	(1)	work content (i.e. actual work done),
	(2)	work method (i.e. manner in which work is executed).
Organization	(3)	work organization (i.e. manner in which work is planned and controlled),
	(4)	job opportunities,
	(5)	social work conditions/relations.

Thirty-six relevant papers, books, etc. were examined. Together the authors of these documents cited thirty-six desirable job characteristics which are categorized in Exhibit 5.3 and examined below.

Work content

Examination of the nature of the fourteen job characteristics listed, reveals a three-level structure, i.e.

I. *Tasks*, i.e.
new and more difficult tasks to be added,
inclusion of some auxiliary and preparatory tasks.

The characteristics listed also provide some examples of the above, i.e.

inspect own work,
repair defects,
set-up machines,
responsibility for cleanliness of work area,
responsibility for maintenance.
II. *Task relationships*
'closure'—perform complete module of work,
obvious relationship between tasks.
III. *Work attributes*
perceivable contribution to product utility,
increased task variety,
use of workers' valued skills and abilities,
meaningful and worthwhile job.

These three levels might be seen to have cause and effect relationships, i.e. it might be advocated that jobs should be structured such that the tasks done bear

JOB/WORK

(1) *Work Content*

    A   'Closure'/i.e. complete module of work
    B   Obvious relationship between tasks
    C   New and more difficult tasks added
    D   Increased variety of tasks
    E   Makes use of workers' valued skills and abilities
    F   Includes some auxiliary and preparatory tasks
    G   Individual inspects own work
    H   Assembler repairs defective items
    I   Operator sets up machines
    J   Operator responsible for cleanliness of work area
    K   Operator responsible for maintenance
    L   Perceived contribution to product's utility
    M   Work content such that job is meaningful and worthwhile

(2) *Work Methods*

    A   No machine pacing

ORGANIZATION

(3) *Work Organization*

    A   Gives worker some choice of method
    B   Worker discretion
    C   Operator plans own work
    D   Operator organizes own work
    E   Self-regulation
    F   Worker responsible for controlling own work
    G   Operator sets own performance goals
    H   Subgoals to measure accomplishment
    I   Individual accountable for own work
    J   Job responsibilities (generally)
    K   Worker autonomy
    L   Operator involved in solving problems
    M   Workers participate in design and improvement of own job
    N   Workers involved in decision making concerning work
    O   Workers receive performance feedback at regular intervals

(4) *Job Opportunity*

    A   More than minimum required training provided
    B   Worker able to learn new things about process
    C   Promotion prospects for worker .
    D   Specific or specialized tasks enable worker to develop expertise
    E   Increased challenge for worker

(5) *Social Conditions/Relations*

    A   Conversation either easy or impossible
    B   Facilitates workers' movement about factory

Exhibit 5.3 Desirable characteristics of jobs

a holistic relationship in order to provide certain work attributes. In terms of authors' advocacy of these characteristics emphasis has been given to the provision of new and different tasks, responsibility for inspection, closure and increased task variety. Few authors cite responsibility for cleanliness and maintenance as desirable job characteristics; however, authors' advocacy of characteristics may be based on their assessments of normal work content (normally includes responsibility for cleanliness?) and practicability (maintenance of equipment?).

Increased worker responsibility is most often quoted as the effect of changes in work content. Recognition, responsibility, achievement and accomplishment are given as the principal effects of task closure and the effective interrelationship of tasks, whilst the addition of tasks and increased task variety is commonly considered to lead to reduced boredom and fatigue, worker involvement, accomplishment, greater use of skills, opportunity for learning and increased job satisfaction.

Work method and organization

Examination of the *work-organization* characteristics reveals a possible two-level structure, namely work organization and work attributes, i.e.

I. *Work organization*, i.e.
  worker has some choice of work method,
  worker plans/organizes own work,
  worker controls own work/self-regulation,
  workers set performance goals,
  regular performance feedback,
  worker participates in job design/improvement,
  worker involved in work problem solving.
II. *Work attributes*, i.e.
  worker discretion/decision making,
  worker accountability/responsibility,
  worker autonomy.

I above may be seen to give rise to the attributes of II and further; although only one item is quoted, some *work-method* characteristics may be seen as prerequisites for certain organizational characteristics (self-regulation is given as one effect of the removal of work-pacing). There is no clear pattern in the suggested relationship of effects and characteristics. Worker involvement is most often given as a consequence of the listed work attributes and several attributes, particularly responsibilities and autonomy, are considered to result from the provision of the other job characteristics. Responsibility and achievement are most often quoted as effects in this section.

A larger proportion of authors cite work-method and organization characteristics than any of the three other groups of characteristics. Furthermore, those referring to such characteristics are in close agreement as to the merits of such characteristics. Responsibility for planning and controlling own work, regular performance feedback, choice of work method and involvement in problem solving receive most attention, whilst, amongst attributes, worker responsibility and autonomy are emphasized.

Job opportunity

Of the five cited characteristics four relate to personal development whilst one relates to job advancement. Worker involvement in the organization (involvement, identity, feelings of importance) and self-actualization (growth and advancement, self-development and pride) are given as consequences of opportunities for worker development, but no specific effects are associated with opportunities for promotion. It could be argued that personal development, e.g. increase in skills, abilities and accomplishment, will give rise to opening employment and hence promotion opportunities, although it is not clear whether authors advocating development characteristics had this consequence in mind.

Social conditions

Both characteristics in this section relate to ease of social interaction. The consensus suggests that a job should facilitate social interaction in the interests of job satisfaction, although it is recognized that complete lack of verbal contact may be preferable to contact with difficulty.

Examination of the information summary and tentative structures developed above suggests a breakdown of the original job characteristics or 'means' category into two sections, i.e. firstly work and job attributes and secondly those characteristics (tasks, task relationships, work methods and organization) which in some combination provide for the existence of such attributes. From this interpretation it appears that it is largely the provision and manipulation of these latter characteristics which gives rise to the existence of work and jobs with attributes capable of causing worker responses. Whilst the processes of job restructuring and work organization do not necessarily follow such a neat path, this three-part process does provide a useful reference framework. The model is summarized in Exhibit 5.4 in which certain additions have been made as a result of the findings cited earlier in this chapter. This model does not necessarily provide a complete checklist, nor does the information reviewed above yield only to this interpretation. However, the structure is of value in that it helps to distinguish between those aspects of jobs which might be manipulated and those job features which changes might affect but which cannot be directly treated. The job-attributes identified here might in many circumstances provide more appropriate objectives for job changes than many of those usually identified in the literature, since these attributes are more easily associated with the 'enabling' job characteristics.

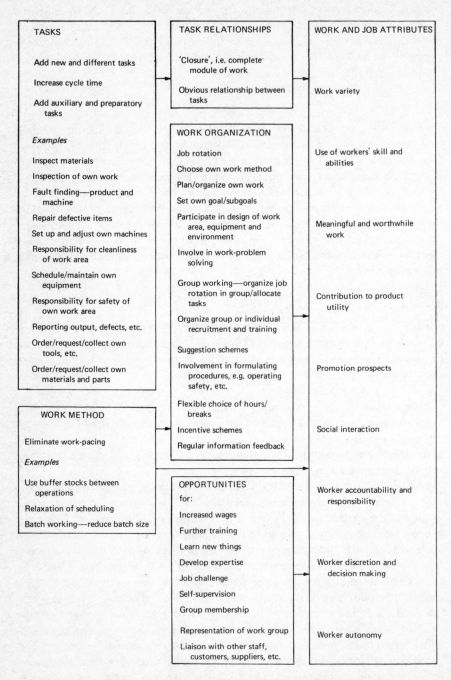

TASKS

Add new and different tasks

Increase cycle time

Add auxiliary and preparatory tasks

Examples

Inspect materials

Inspection of own work

Fault finding—product and machine

Repair defective items

Set up and adjust own machines

Responsibility for cleanliness of work area

Schedule/maintain own equipment

Responsibility for safety of own work area

Reporting output, defects, etc.

Order/request/collect own tools, etc.

Order/request/collect own materials and parts

WORK METHOD

Eliminate work-pacing

Examples

Use buffer stocks between operations

Relaxation of scheduling

Batch working—reduce batch size

TASK RELATIONSHIPS

'Closure', i.e. complete module of work

Obvious relationship between tasks

WORK ORGANIZATION

Job rotation

Choose own work method

Plan/organize own work

Set own goal/subgoals

Participate in design of work area, equipment and environment

Involve in work-problem solving

Group working—organize job rotation in group/allocate tasks

Organize group or individual recruitment and training

Suggestion schemes

Involvement in formulating procedures, e.g. operating safety, etc.

Flexible choice of hours/breaks

Incentive schemes

Regular information feedback

OPPORTUNITIES

for:

Increased wages

Further training

Learn new things

Develop expertise

Job challenge

Self-supervision

Group membership

Representation of work group

Liaison with other staff, customers, suppliers, etc.

WORK AND JOB ATTRIBUTES

Work variety

Use of workers' skill and abilities

Meaningful and worthwhile work

Contribution to product utility

Promotion prospects

Social interaction

Worker accountability and responsibility

Worker discretion and decision making

Worker autonomy

Exhibit 5.4 Model relating the characteristics and attributes of work and jobs

References

1. Maslow, A. H., 'A theory of human motivation', *Pyschological Review*, **50** (1943).

2. Herzberg, F., *Work and the Nature of Man*, World Publishing Co., 1966.

3. Reif, W. E., and Schoderbek, P. P., 'Job enlargement: an antidote to apathy', *Management of Personnel Quarterly*, **5**, 1 (1966), pp. 16–23.

4. Wilkinson, A., *A Survey of Some Western European Experiments in Motivation*, Institute of Work Study Practitioners, 1970.

5. *Work Structuring: A Summary of Experiments at NV Philips*, Philips, Eindhoven, 1969.

6. Whyte, W. F., *Money and Motivation*, Harper, 1955.

7. Anderson, J. W., 'The impact of technology on job enrichment', *Personnel*, **47**, 5 (1970), pp. 29–37.

8. Mann, F. C., and Hoffman, R. J., *Automation and the Worker*, Holt, Rinehart and Winston, 1960.

9. Ford, R. N., *Motivation through the Work Itself*, A.M. Association, 1969.

10. Weed, E. D., 'Job enrichment "cleans up" at Texas Instruments', in Maher (Ed.), *New Perspectives in Job Enrichment*, Van Nostrand Reinhold, 1971.

11. Doyle, F. P., 'Job enrichment plus O.D.' in Maher (Ed.), *New Perspectives in Job Enrichment*, Van Nostrand Reinhold, 1971.

12. Maher, J. R., *et al.*, 'Enriched jobs mean better inspection performance', *Industrial Engineering* (November 1969), pp. 23–26.

13. Paul, W. J., *et al.*, 'Job enrichment pays off', *Harvard Business Review* (March–April 1969), pp. 61–78.

14. Vroom, V. H., *Work and Motivation*, Wiley, 1964.

15. Dickson, J. W., 'What's in a job', *Personnel Management*, **3**, 6 (1971), pp. 38–40.

16. Little, A., and Warr, P., 'Who's afraid of job enrichment?', *Personnel Management*, **3** (1971).

17. Maher, J. R., *New Perspectives in Job Enrichment*, Van Nostrand Reinhold, 1971.

18. Powers, J. E., 'Job enrichment: how one company overcame the obstacles', *Personnel* (May–June 1962).

19. Chaney, F. B., 'Employee participation in manufacturing job design', *Human Factors*, **11**, 2 (1969), pp. 101–106.

20. Sirota, D., and Wolfson, A. D., 'Job enrichment: surmounting the obstacles', *Personnel* (July–August 1972).

21. Smith, D. M., 'Job design: from research to application', *Journal of Industrial Engineering*, **XIX**, 10 (1968), pp. 477–482.

22. Colosky, D. A., 'How to waste and misuse human capabilities', *Assembly Engineering* (July 1972).

Part 3
A Review of Some Recent Cases

CHAPTER 6

Summary and Review of Cases

Appendix B contains seventeen very brief case descriptions drawn from eleven European companies. They describe changes which have taken place, or new methods employed, within these companies, and have been selected as examples from a larger series of thirty-seven case studies relating to these and other companies in Europe. Details are given in Exhibit 6.1. These thirty-seven cases were prepared after company visits during 1972–1973 and together appear to represent a reasonable cross-section of developments taking place in discrete item mass production around that period.

Limitations on space prevent the presentation of details on all cases, thus those appearing in abbreviated form in Appendix B should be taken solely as examples. They relate to four industrial sectors—vehicle production, mechanical engineering, electrical engineering and electronics and domestic and office appliances; however, they do not themselves represent an accurate sample within these sectors nor should the cases be compared without reference to the circumstances which applied to each. The descriptions are presented in largely factual terms, without judgement or evaluation. Anonymity is preserved excepting where information has been previously published.

We shall use this case material, or, more importantly, observations on these thirty-seven cases, in order to test and extend some of the concepts developed in Chapters 4 and 5. The discussion which follows relies upon observations on all cases listed in Exhibit 6.1 and therefore constitutes a form of commentary on recent developments in mass production in these sectors in Europe in recent years.

Reference will be made to *changes* throughout the discussion; however, it should again be emphasized that although most cases deal with changes in, or modifications of, previous practice, new situations are neither excluded nor identified separately. Thus a 'change' should be seen as an example of development in this aspect of mass production, whether a new method or system (as in cases A, B, C, D, H, N) or a modification or development of previous practice (as in cases E, F, G, J, K, L, M).

SUMMARY OF NATURE OF CHANGES

The majority of the changes observed dealt either with the introduction of individual assembly (in 30 per cent of the cases) or organizational changes

63

Case	Subject	Company	Country
Motor-Vehicle Manufacture			
1	Truck manufacture	A	Sweden
B*	Engine assembly	A	Sweden
2	Car assembly	A	Sweden
A*	Motor-vehicle production	B	Sweden
C*	Gearbox assembly	C	U.K.
D*	Engine assembly	D	U.K.
Mechanical Engineering			
E*	Manufacturers of vehicle parts	J	Norway
3	Assembly of handsaws	K	U.K.
4	Coal-fire assembly	L	U.K.
5	Convector assembly	L	U.K.
6	Cabinet assembly	L	U.K.
7	Control-unit assembly	L	U.K.
Domestic and Office Equipment			
G*	Floor-sweeper assembly	Q	U.K.
8	Floor-mop assembly	Q	U.K.
9	Assembly of bells and chimes	R	U.K.
10	Assembly of vacuum cleaners	S	U.K.
H*	Washing-machine assembly	T	U.K.
11	Washing-machine assembly	T	U.K.
J*	Steam-iron assembly	U	U.K.
12	Domestic-appliance manufacture	V	Sweden
13	Tape-recorder assembly	W	Norway
K*	Television assembly	Y	U.K.
14	Television remote-control manufacture	T	Holland
L*	Television assembly	T	Holland
M*	Typewriter assembly	Z	Holland
15	Testing washing-machines	T	U.K.
16	Assembly of hi-fi equipment	X	U.K.
Electrical and Electronics			
17	Assembly of registers	AA	U.K.
18	Assembly of electric meters	AA	U.K.
19	Assembly of switchgear	BB	Norway
20	Assembly of television and electric equipment	T	Denmark
21	Assembly of meters and switches	CC	Denmark
22	Assembly of relays, etc.	DD	Norway
23	Finishing of television valves	EE	U.K.
24	Valve testing	EE	U.K.
25	Supervision of valve manufacture	EE	U.K.
F*	Valve assembly	EE	U.K.

* Cases appear in abbreviated form in Appendix B.

Exhibit 6.1 Case examples

involving some degree of worker self-organization (45 per cent). In the latter cases the self-organization facility was based almost entirely upon the creation of some form of group working in which some, often considerable, degree of freedom was offered to the workers. Such self-organization in general included the opportunity for job rotation. Excluding those changes which were introduced as part of, or as a consequence of, worker self-organization, the changes observed also covered the use of joint consultative groups, job enlargement through reduction of flow-line length, reduction or elimination of mechanical pacing, increased worker responsibility and further rationalization of tasks through increased mechanization.

It is interesting to note that many of the more comprehensive changes, particularly those involving organizational changes, occurred in companies outside the U.K. This applies in all four sectors, indeed in general changes in the U.K. were generally far narrower and were more likely to have been influenced by the need for operational improvements, e.g. flexibility, quality, productivity, etc. Changes in the U.K. were, it seems, more technique-oriented and were in general concerned with (or had the effect of giving rise to) job enlargement, limited job enrichment and job rotation, whereas changes, in many cases in similar situations, abroad tended to deal more with vertical enrichment and worker self-organization. Equally, however, it is clear that the changes in these latter cases were more likely to have resulted from policy decisions, were more likely to have been part of a wider programme of change, which in turn was likely to have been influenced by both remedial and preventative considerations based in the main upon an assessment of the behavioural characteristics of mass-production jobs. In fact such changes appeared to have resulted for reasons which closely resemble the now classical arguments in favour of job restructuring, whilst most of the changes in the U.K. appeared to have been brought about by engineering and operational considerations.

With the exception of the two Swedish vehicle manufacturers, few changes were evident in heavy-goods mass production. The changes that were proposed in this sector, however, were in general of a far larger scale and scope than those found in technologically similar situations in the mechanical-engineering sector. Indeed in mechanical engineering very few changes were evident. In contrast, the two remaining sectors provide a wide range of situations and a variety of changes and systems. In particular the manufacture of domestic, consumer and office equipment provided good examples of virtually the whole range of changes identified in the entire study.

JOB RESTRUCTURING AND WORK ORGANIZATION

The changes taking place in recent years in discrete-item mass production cover an extremely wide range differing in detail, scope, orientation and even basic philosophy. Using the terminology developed in Chapter 5 it is evident that virtually all of the major exercises fall into the category of organizational change embracing both worker self-organization and job rotation—particularly the former. (78 per cent of the developments studied in companies in Scandinavia and Holland and 70 per cent of the companies visited provided examples of such organizational change.) In virtually all of these cases the changes have provided a basis for job restructuring which in turn generally provided some of the characteristics conventionally associated with job enrichment.

Self-organization formed a major part of most organizational changes. In general the facility for self-organization was afforded to groups of workers rather than to individuals, and in many cases the creation of such *group working*' was apparently the principal, even the sole, objective. This concept of 'group working' featured as a major issue in all four sectors and provided one of the major themes of recent work organizational changes. (Group working will be examined in detail in Chapters 7–10.) In comparison to the changes effected in the nature of work and jobs, either as a consequence of or as part of an organizational change, work and job changes introduced separately were generally more limited in scope. They normally afforded job enlargement and generally resulted from the manipulation of flow-line parameters, e.g. the reduction in line length or the introduction of individual working. The motives behind many of these changes provide the second important theme emerging from the studies, namely restructuring–enlargement, and exceptionally the enrichment of work, for operational rather than behavioural reasons, i.e. restructuring as a consequence of changes undertaken for other reasons.

The means and ends of job restructuring

Reference to the model developed in Chapter 5 (Exhibit 5.4) facilitates further evaluation of these changes.

If we look at the changes which primarily involved the rearrangement of flow lines, i.e. the introduction of shorter lines (Case G), and ultimately the adoption of individual assembly (e.g. Cases C and J), we see that such job restructuring is provided in many cases by task changes and, to a lesser extent, changes in work method.

In most cases workers were required to undertake some of the work which would otherwise have been undertaken by a colleague. Thus increased work variety was provided mainly where workers were required to undertake some of the auxiliary and preparatory tasks which might otherwise have been undertaken at one part of a flow line. For example, the use of individual assembly often necessitated some inspection and test work to be undertaken by each assembler. Auxiliary and preparation tasks were often provided in that workers were also required to rectify items subsequently found to be incorrectly assembled. These additional tasks, together with the increased direct work, might well relate together in such a manner as to give rise to task closure, which in turn may help provide jobs with the attributes of meaningful and worthwhile work and perceived contribution to product utility. To this extent, such job restructuring satisfied some of the requirements identified in Chapter 5. This limited effect, however, appears to reflect both the motives and the means employed. The motives generally related to the desire for operational improvements, such as reduced balancing loss, improved flexibility and improved quality, whilst the means for restructuring relied upon the redistribution of the total work content of the job, without the addition of different tasks. The behavioural limitations of such changes are well illustrated by the method of individual assembly adopted in Case C which provided machine-paced work having an 'increased' cycle time but without the addition of any auxiliary or preparatory tasks. This system perhaps afforded some closure, but equally it also provided probably the best example of a 'change' which was unacceptable to most of the people involved, and provided few of the behavioural benefits normally associated with job restructuring.

Similar restructuring changes were found to have taken place in entirely different circumstances. For example, in Case N workers were each required to assemble a complete item. Indeed in some respects task variety was low since workers were not required to undertake their own inspection. The difference between this and the examples above, however, is to be found in the nature of the work organization. A similar example was found in Case B. The system for engine assembly adopted in this company provides for individual assembly or a form of flow-line working. Workers undertake few of the additional auxiliary or preparatory tasks listed in Exhibit 5.4, indeed in this respect when working individually they differ little from the workers engaged on gearbox assembly in Case C, excepting in having a far larger cycle time. Again, however, in this system workers were able, as a group, to choose their work system, allocate tasks, organize the job, etc.; thus in terms of required job attributes their jobs may be considered to be superior to those referred to above. Similar differences in approach are evident in several other cases, for example the new method of television assembly adopted in Case K involved the restructuring of flow-line work through the reduction of line length together with an increase in cycle time and an increase in interstation buffers. Few additional tasks were allocated to workers. In contrast, television assembly in Case L provides the same assembly system within a redesigned organizational context; hence workers' jobs are enriched largely through the provision of tasks which were not previously the responsibility of assembly workers. In the former case the main motive for the change was increased production flexibility.

Further examination of the model developed in Chapter 5 suggests that only a small number of the required attributes of restructured jobs may be provided by task changes. Examined against the list of required job attributes given in Exhibit 5.4, the following cases from Appendix B would appear to be amongst those which most closely satisfy the behaviour requirements of job restructuring.

Case B	Engine assembly	(Sweden)
Case A	Motor-vehicle assembly—(Kalmar plant)	(Sweden)

Case E	Machining of brackets for brakes	(Norway)
Case F	Valve assembly	(U.K.)
Case H	Washing-machine-drum manufacture	(U.K.)
Case N	Assembly of remote control	(Holland)
Case L	Assembly of television	(Holland)
Case M	Typewriter assembly	(Holland)

In each case the systems provided jobs which possess many of the attributes identified in Chapter 5. Meaningful and worthwhile work, work variety, perceived contribution to product utility, worker discretion and decision making, accountability, responsibility and autonomy may all exist. However, it is the possession of the attributes at the end of this list that sets these cases apart. Unlike many other cases where restructuring resulted from the redistribution of the existing work content of the jobs, these cases involve some vertical integration of tasks. Thus, for example, certain tasks which were previously undertaken by supervisors, quality controllers, materials handlers and industrial engineers are allocated to the workers, who as a result become to some degree self-organizing. In many cases the reasons for changes, or the objectives of the new system, appeared to be the provision of such self-organization rather than simply the restructuring of jobs.

Limitations on, and effects of, job restructuring

In virtually all cases it was clear that the restructuring of jobs and work reorganization gave rise to far-reaching effects at supervisory levels, and in many cases such exercises obviated the need for the lowest level of supervision. In some cases, the jobs in other staff and technical functions were affected, either directly—following the allocation of certain responsibilities to the workers—or indirectly, for example by necessitating direct contact with workers rather than their supervisor. In some cases the effects of job restructuring and work reorganization on other personnel were the main threat or limitation to the success of the exercise.

In most organizational changes the added responsibilities given to workers were transferred partly from supervisors; thus supervisors whose duties were largely repetitive rather than technical were, on occasions, no longer required, e.g. the role of chargehands might be eliminated (e.g. Case L). Those remaining supervisors were often required to devote more time to technical duties and less to direct work involvement on the shop floor, which in turn often necessitated certain extra training for supervisors.

A surprising number of job-restructuring exercises were resisted by workers and in some cases changes had to be abandoned or modified as a result of such resistance. Some resistance appeared to derive from disagreement over wage rates and production targets, and clearly the nature of the payment system, the wage grading of employees and the determination of production standards are crucial issues in the restructuring of jobs. In many cases workers whose jobs have been restructured either received or pressed for increases in wages, normally in the form of wage-grade increases, to compensate for their utilization of further skills, their flexibility or assumption of greater responsibility. In many cases wage increases had been given as a prerequisite of the change and in one case (E) the change was sought in order to provide increased pay. Since in many cases the introduction of organizational changes led directly to the reduction in supervisory requirements, and therefore in the costs of supervision, many companies had been prepared to offer increased financial rewards to workers. In contrast, simple job enlargement was often not seen as ground for increased wages; even so pressures for pay increases in such situations occasionally jeopardized the success of the new methods (Case C).

In many situations responsibilities and tasks were shared by workers and hence differential wage grades were considered to provide a barrier to cooperation and to success. Many of the companies that had introduced such systems employed a day-work system of payment (Cases L, M, N) whilst in certain cases, it was also considered to be important that the wage system employed should reflect, and preferably emphasize, group interdependence; hence individual incentive systems were considered to be dysfunctional, some form of group-payment system being desirable. Furthermore, since the integration of tasks often implied the allocation of both direct and indirect responsibilities to workers, it was considered desirable that if an incentive payment system was employed it should reflect the indirect responsibilities of workers (e.g. the tooling allowance in Case E).

Further training for operators was provided in some, but by no means all, cases. It appears that the provision of extra training is not always necessary in order to achieve required performance levels on restructured jobs. This may in some cases be offset by somewhat different training needs for both workers and supervisory/technical staffs. This is particularly evident when some form of organizational change is introduced, since in such cases workers will generally be required to assume greater responsibility, interact differently and fulfil different roles. Such situations also appear to have a major effect upon the roles of supervisors and auxiliary and technical staffs.

Extra space requirement is often advanced as one disadvantage of restructured jobs, particularly where such restructuring leads to the abandonment of rationalized flow-line work. Several of the studies, however, provide evidence which conflicts directly with this contention; indeed in some cases changes had been made partly to save space. Equally, whilst objective assessment was not usually possible, it was evident that many of the revised methods did give rise to inferior space- and equipment-utilization, particularly in situations in which workers possessed considerable freedom, since such conditions generally necessitated the provision of spare work benches, tools, etc. It seems, therefore, that the abandonment or modification of conventional flow-line working might in principle lead to the need for increased investment in space and equipment; such an effect does not necessarily apply in practice, especially when an existing line providing for large work in progress and material stocks is to be replaced.

Only 40 per cent of the cases examined appeared to relate to situations in which some form of remedial action had been taken. For example, changes to overcome problems of absenteeism, quality, etc. A form of preventative action had been taken in 15 per cent of cases, e.g. changes designed simply to increase job satisfaction, etc.

In very few cases had changes been evaluated in any detail. Company A had undertaken a detailed cost analysis of their new method for engine assembly (Case B) although it had been possible only to provide estimates of certain of the costs used in this evaluation. Company T (Case L) had undertaken what was undoubtedly the most detailed evaluation, although in this case the changes had been introduced experimentally, with the objective of permitting such an evaluation. In both cases the production systems, which provide both organizational and structuring changes, were found to offer economic gains, although in neither case was the benefit substantial. In the latter case the evaluation also included a detailed assessment of the behavioural aspects of the new system, and again an improved situation was found to have been obtained.

In general the effects of the restructuring changes were found to have been evaluated simply in terms of the reasons for the changes, i.e. whether or not the objectives had been achieved. Thus in those cases where simple job changes had been introduced for operational reasons, subsequent evaluation (which was rarely quantitative) concentrated on issues such as flexibility and quality, little attention being given to 'indirect' or secondary effects.

AUTOMATION AND MECHANIZATION

Applying the definitions developed in Chapter 4, virtually all examples of technological change observed in this aspect of mass production could be described as increases in the level of mechanization. Very few examples of automatic control beyond level 2 (Exhibit 4.1) were to be found; hence in most cases workers were responsible for monitoring and other control duties. Furthermore, many of the systems examined also required manual intervention in either 'processing' and/or 'handling'; hence workers were required to perform both direct and indirect production activities.

Both metal-working and assembly systems were examined. In the former (e.g. Case E) all processing was provided by automatic-cycle, hand-activated machine (i.e. level 4—Exhibit 4.1). Handling between machines was provided by conveyor systems (level 3) machine loading/unloading being performed almost entirely manually, (e.g. levels 2 and 3). Assembly systems, with the exception of printed-circuit-board assembly, typically provided for 'processing' automation, to level 4 or 5, whilst manual handling was generally retained in either feeding items to machines or to hopper and automatic-feed devices. Thus the pattern of development identified in Chapter 6 was evident, i.e. higher levels or earlier development of automation in processing for metal-working systems, and in handling for assembly systems.

Examples of each of the three types of assembly work identified in Exhibit 4.2 were to be found, i.e. preparation and component manufacture, subassembly manufacture and final assembly. Preparation and component manufacture, in for example printed-circuit-board assembly, was generally automated to stage 2, i.e. subassembly and final assembly was generally accomplished by systems which employed some degree of mechanized handling together with manual assembly (i.e. also stage 2). In terms of comparatively automated systems automation in both cases was evident at stage 3, i.e. mechanized processing and handling for subassemblies and mechanized handling with some process mechanization for final assembly.

By far the largest number of systems used were employed for subassembly manufacture. In general such systems utilized single-station or rotary-indexing assembly machines, fed and monitored by manual operators. The more advanced systems provided for the linking of such machines, either by conveyors to hopper feeds or manual-load operations, or by pick-and-place devices. Multi-station in-line indexing machines were evident only in metal working and in the assembly of comparatively heavy and bulky items (e.g. engine assembly in Case D).

Virtually all examples of comparatively highly automated assembly work was found in the electronics and elecrical-engineering sector. This sector provided many examples of the use of both single-station and rotary-indexing machines as well as 'linked' machines. Indeed in many cases there was evidence that the development of automation in this sector involved firstly the development and 'debugging' of operator-controlled and relatively simple single-station or rotary machines for component or subassembly work, followed by the integration of such machines to form a linked line, the linkages being provided initially by manual-loading operations, but ultimately by some form of mechanized loading. Since these machines were generally employed for the assembly of small items, substantial buffer stocks normally existed between machines to protect against machine stoppages. Thus, although these configurations might be considered to be non-synchronous indexing systems, their manner of operation had more in common with linked machines lines.

Reasons for and results of automation/mechanization changes

The principal reasons given for efforts to increase levels of automation were:

The need to improve/ensure product quality.
The need to reduce product costs.
The need to increase output/productivity.

The desire for quality improvements was particularly evident in the electronics industry. Here it was repeatedly demonstrated that customer quality and performance standards were continually increasing, and this together with the continuing miniaturization of items and products was said to lead necessarily to efforts to eliminate manual-assembly tasks. This trend was especially evident in the telecommunications industries, where customer standards on, for example, the quality and reliability of soldered connections were becoming tighter. In many cases in the telecommunications industry output quantities of certain major items, e.g. printed-circuit boards, were small and product diversity large (e.g. several hundred different circuit boards are required for one rack for a telephone exchange). Nevertheless, considerable efforts were being made to increase the level of automation in order to satisfy the demands made on the company.

In other situations within these same industries items were made in very large quantities, e.g. items for telephone handsets. In such cases automation was generally seen as the only means to increase output and reduce item costs. Typically, in these cases product value was small, the range limited and product life easily predicted.

In few cases was there any suggestion that efforts to increase the degree of automation were connected with the desire to eliminate jobs which were, or might become, unacceptable to workers. Hence, whilst automation has been identified as one possible solution to such behavioural problems this motive was not in evidence during this study. However these two issues were found to be closely associated, since many of the behavioural consequences of partial automation identified in Chapter 6 were evident in the systems examined in the study.

Work roles and assembly automation

It was evident, in virtually all cases, that whilst the use of further automation generally improved labour productivity and therefore reduced labour requirements, such systems did not operate independently of manual intervention. The continuing need for direct operator movement was evident in all cases of increased assembly automation, and in all such cases the work undertaken by workers appeared to provide low variety, and little scope for discretion (Exhibit 4.3). In all situations where process automation had been developed to level 4 or beyond (Exhibit 4.1), the remaining manual tasks provided short cycle times—often less than ten seconds, together with some degree of mechanical pacing. Furthermore, even in situations where direct manual processing (i.e. assembly) work was retained, the role of the worker was impoverished in comparison to that which might have existed at the prior stage of system development.

In no case were workers required to perform tasks above the level of machine minding identified in Chapter 4 (Exhibit 4.3), i.e. the predominant roles of workers in these systems were 'in-line and auxiliary work' and, to a lesser extent, machine minding. Furthermore, it was judged that such roles, particularly in-line and auxiliary work, provided the job attributes discussed in Chapter 4, i.e. low variety, little discretion/responsibility and little perceived contribution. Not only did these attributes appear to distinguish these jobs from those that might exist at higher levels of automation, but also they were considered to be evident to a greater or further degree than might have been the case for, say, manual flow-line working. Additionally these characteristics were perhaps accentuated by increased social isolation, although the physical effort required of the worker was generally low. Machine-minding work was occasionally linked with auxiliary work such as loading/unloading, when the latter was undertaken without close machine pacing (e.g. hopper loading, etc.). In such cases the workers' predominant role appeared to benefit in terms of work variety, discretion and perceived contribution. However, in only one case were such duties a dominant part of the individual's role.

Thus the general pattern identified in Chapter 4, is supported by observations during the study, and certainly few, if any, of the behavioural benefits associated with the higher levels of automation common in the process industries were evident in assembly automation. For this reason assembly automation and mechanization should perhaps be considered alongside job restructuring, even though the hypothesis of Chapter 3 is not borne out by the observations made during this study.

Part 4

The Concept of Group Working

Work groups, their nature, design and significance, and the nature of group working.*

*Readers familiar with the concepts and theories of work groups may wish to omit parts of Chapters 7 and 8.

CHAPTER 7

Group Working: The Nature and Types of Work Groups

The importance of group working in the development of mass-production work is readily evident from the review presented in the previous chapter. A large proportion of recent cases of changed or changing methods have been found to some degree to concern what has been referred to as *group working*. Such efforts are particularly evident in Scandinavia, indeed the work undertaken in Norway, and latterly in Sweden, has recently attracted a great deal of attention in other countries[1,2] for this reason.

The evident importance of this concept in mass production is clear, if only from the amount of attention that has been focused on it. However, despite, or perhaps because of, this considerable interest in the subject, the precise nature of group working remains unclear. The fact that many of the examples quoted in the literature relate to diverse situations (ranging from logging to heavy engineering) has perhaps helped demonstrate the scope of potential applications, but as yet there appears to be no detailed assessment of actual or potential application in one particular industry.

In order to provide a sufficiently comprehensive and self-contained reference source for what appears to be a particularly important issue, and in order to provide a basis from which to examine the application of group working, two chapters are devoted to the examination of the nature, characteristics, benefits and design of group working. The primary objective in these chapters is, by reference to relevant and often fundamental information, to identify the meaning of the term group working prior to examining applications in engineering mass production.

This chapter will examine the nature and types of group working. In the following chapters the benefits of group working will be explored and some basic design factors will be identified.

GROUP WORKING AND PRODUCTION SYSTEMS

Individual working, as defined in Chapter 2, suggests the absence of significant division of labour and product flow. The term is useful in a comparative sense insomuch as it may conveniently describe a system in which each person assembles an item which might otherwise be assembled progressively on a flow line. Any degree of job enlargement of flow-line working might be considered to lead to individual working, but here the term implies the assembly of a significant item. In some senses individual working may be thought to be similar to the now largely obsolete methods of manual quantity production, and whilst historically much manual assembly work has been undertaken by individual workers this has often been associated with either low-quantity output or simple products.

If we view individual and flow-line working as contrasting systems of manual assembly, then it is possible to consider group working as representing an intermediate system. Group working might therefore be seen as a system to replace progressive flow-line working without necessitating individual working, since in group working the assembly of items is accomplished collectively by a group of workers rather than by an individual.

The manner in which a group of workers might be employed in the assembly of items is explored in Exhibit 7.1. A simple case of two-man group is taken,

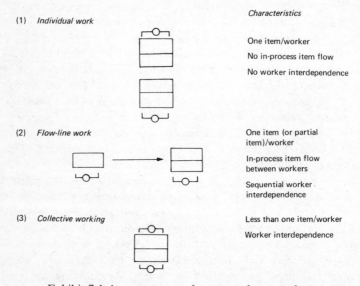

Exhibit 7.1 Arrangements of a group of two workers

and three basic systems are shown, each of which provides the same output.* Although this gives an oversimplified view it is sufficient to indicate that group working can embrace both individual (1) and flow line arrangements (2), whilst possibly providing a third system (3). This latter resembles flow-line working in that in a steady-state condition each worker undertakes a part of the total assembly work content, yet it differs in requiring less than one item in progress per operator. In order to avoid confusion we shall refer to this system

* Ignoring losses through inbalance and interference.

as *collective working* whilst for methods (1) and (2) respectively we will retain the terms individual and flow-line working.

In practice collective working is often employed at stations on flow lines used in the manufacture of large items such as cars. Such work systems, and various types of collective working, will be examined later; however, for the present purposes it is sufficient to note than an examination of work and production systems reveals very little about the nature of group working as dealt with in much of the recent literature on mass production, since all such systems normally involve a group of workers. We shall examine the nature of work systems and their relationships with work groups later, whilst in this chapter we will continue to seek information relating to 'group working', firstly through an examination and review of the socio/psychological theory of *work groups,* their formation, characteristics and dynamics, and then through a study of the *group-working experiments* and exercises undertaken in industry.

WORK GROUPS—NATURE, TYPES AND FUNCTIONS

Three basic types of worker activity might be observed in most production situations. The principal activity derives directly from the nature of the work content of the task, which places certain demands on the skills of the worker. Exercise of such skills constitutes a form of *technical activity.* The demands of the work-place, and the nature of the work to be done, also necessitates some interaction amongst workers, thus a second type of behaviour—here referred to as *socio-technical*—is required, and includes any form of social interaction necessary for the performance of the task. The third level of behaviour is purely *social.*

These three basic types of activity are interrelated, hence technical organization and socio-technical activity must be examined in order to understand worker behaviour. For example, the formation of many types of work groups is considerably affected by the technical and formal organization of the work place.

Types of work groups—a summary

Any number of workers sharing certain characteristics and relating one to another in such a way as to differentiate them from others might be considered to constitute a work group. A variety of situations in industry provide for the formation of such groups, e.g. the work group sharing a common objective of the completion of a certain task, the 'card-school' or the collection of workers supervised by one foreman. Such groups have in common the fact that some form of interaction takes place between their members.

Cartwright and Zander[3] have suggested that the relations between members must be such that they become interdependent to some significant degree; similarly Lewin[4] considers that group formation is the result of interdependence, whilst Schein[5] has noted that there is a difficulty in defining a group independently of some specific purpose, or frame of reference.

Groups within organizations may be categorized in several ways, the principal distinction being as follows:

(1) *Formal groups*, created to achieve specific goals and to carry out specified tasks which are clearly related to the total organizational mission. They may be either permanent or temporary groups, depending upon the purpose of their formation.

(2) *Informal groups*. Relationships will develop between members of the organization which extend beyond functional objectives. If the arrangement of the work area, the work schedule and the nature of the work permit, these informal relationships may lead to the development of 'informal groups'. Such groups arise out of a particular combination of 'formal' factors and human needs and are affected by, among other factors, the degree of interaction between individuals, personal characteristics, interests and external influences.

Sayles[6] subdivides these two basic types of groups as follows:

(1) Formal:
   (a) Subordinate group—members share a common supervisor.
   (b) Functional group—members must collaborate in order to accomplish the task.
(2) Informal:
   (a) Friendship clique—members gain certain satisfaction from their interactions.
   (b) Interest group—employees who share a common economic interest and who are held together by their desire to gain common objectives.

Taking the division a little further, *formal functional* groups can be classified by the type of dependence between members, i.e.

(1) Operational interdependence, in which members are dependent for completion of their task upon other members. Groups working in progressive manufacturing systems such as assembly lines are operationally interdependent. Earnings often depend upon the collective performance of the group, in which case pressures are often exerted by the group upon those deviating from group norms. When these groups are found at the shop-floor level there is usually status homogeneity among members.

(2) Functional interdependence, in which members are dependent upon others with complementary skills in order to achieve completion of the group's objectives. This category includes such groups as those manning process equipment and maintenance teams. Individual earnings may be related to the group productivity, and the different skills possessed by group members may lead to status differentials.

(3) Structural interdependence resulting from organizational design, giving members common supervision and normally common territory. Payment may relate to group or individual effort, and members may be heterogenous in respect of skills and responsibilities.

Dubin,[7] examining such formal or organized groups, identifies

(i) Team Group, in which members designate the positions to be filled and the people to fill them, changing allocation of members to positions as required. Such groups are fairly autonomous, receiving very little supervision, and often work within broad terms of reference established by supervision.

(ii) Task Group, in which the jobs are clearly defined and each individual is assigned to one and only one job. The group will have some flexibility over the method of work adopted and also the rate of work, but little other discretion.

(iii) Technological Group, in which work content and method are specified and individuals are assigned to the jobs. Speed of working is also controlled. The individual has little scope for the use of discretion. He has, however, in most cases the opportunity for some degree of social interaction. The overriding importance of the technology allows the group members little autonomy in determining or varying the operating activities.

The types of work group itemized above are included in the approximate relationship given in Exhibit 7.2, which although containing only a small proportion of the types of group identified in the literature, illustrates the essentially two-part structure of group working in industry.

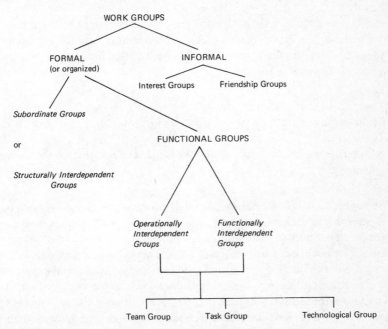

Exhibit 7.2 Summary of types of work group (emphasis being placed on formal groups)

Functions of work groups

Formal work groups are created in industry, whilst informal groups will always exist. As the two types are not mutually exclusive, the functions of groups must be examined with respect to both formal and informal groupings. Individuals are rarely restricted to membership of one group, e.g. two individuals working together may form a group which itself is part of a larger work group, they may independently belong to different social groups, or to formal groups such as a Works Committee. Dual or multiple group membership is important in respect of work behaviour insomuch as some source of conflict may result from the obligations of membership or the objectives of the groups.

A primary function of organization is to achieve goals which are beyond the capability of the individual working alone. Work groups, particularly formal groups, facilitate the subdivision of tasks in order to fulfil formal organizational functions. In addition to satisfaction of economic objectives, groups offer members opportunity for social and psychological satisfaction. Affiliation needs may be satisfied through acceptance by a group, which may further offer individuals opportunity for the development or enhancement of a sense of identity and self-esteem. Members may acquire the opportunity for exercise of influence, and can establish and test reality through the development of consensus amongst members. The effectiveness of social skills can be tested and improved, whilst the group may provide the individual with reinforcement and security.

Tannenbaum[8] has suggested that the support offered to the individual by the formal or informal group is an especially important basis for the member's attraction to the group in the context of the frustration individuals face in their jobs. Such support may be of several types:

(1) Comfort or consolation to members.
(2) Help or protection to members by acting against the source of threat or frustration.
(3) Strengthening the individual member in his opposition to the source of adversity.

The degree to which the behaviour of an informal group corresponds to that required for the achievement of organizational goals depends upon the particular mix of human, technical and organizational inputs to the system. The informal group will generate informal activity aimed at increasing its influence over its environment. There is competition in any organizational environment for scarce resources such as budgetary allocations, space, etc., and any group which fails to obtain an equitable share for its members is likely to disperse, no matter how well it provides social rewards. Without the group's members feeling a sense of social support from the other members, the group would be unable to exert any concerted pressure against the environment.

Group 'rules' are developed in most types of functional work group, especially rules relating to the objectives and important aspects of behaviour of the group, e.g. output, work methods, attitudes to management and Unions and attitudes on other work subjects, and social activities. Argyle[9] summarizes—'Norms are created as a solution to the external problems of the group (the work and environment to be dealt with) and as a solution to the internal problems (how to survive as a harmonious social group).' Such formulation of rules and standards often results in groups becoming conservative and resistant to change. Highly cohesive cliquish groups (see next chapter) may exhibit hostility to other groups and to external authority and this, together with norms relating to group performance, perhaps represents the principal forms of group counter-productive behaviour.

Discussion

The brief review above suggests that work groups of various types pervade industry. Clearly, irrespective of the type of production system adopted, work groups will exist and possibly exert significant pressures which may be counter- or pro-productive. The socio-psychological consequences of work-group membership for the individual are complex, yet it seems clear that for the individual the existence of both formal and informal groups is largely beneficial. Equally, it is clear that the nature of both classes of work group and the existence of formal groups are profoundly affected by the technology and organization of the work, and that those responsible for production-systems design are in a powerful position to influence the nature of work-group relations at shop-floor level. In the following section we shall attempt to identify the nature and purpose of the work groups that have been created through manipulation of the production system in industry. Since we are concerned with the design and comparison of alternative production systems, we are primarily concerned with formal work groups.

SURVEY OF EXERCISES IN GROUP WORKING

In order to learn more about the nature of group working in industry, and to try to place the foregoing discussion in a practical context, a survey of group-working exercises was conducted. The table in Appendix C summarizes the details collected during this survey of published material. The information was gathered from recently published literature and relates solely to industrial exercises or experiments in which some form of group working or group activity was introduced. Only cases in which this form of organization is specifically mentioned and described have been included, since the purpose of the survey was to explore the meaning and nature of group working rather than to attempt to identify elements of group work in any type of changed situation. The survey was conducted for this purpose in preference to the examination of the studies cited in Chapter 6 since the latter had been selected and interpreted by the author, and might therefore present a biased or narrow view. Since the main purpose of the survey was the development of a basic understanding, it was considered to be appropriate to include information relating to any type of 'blue-collar' production activity, whether or not in engineering mass production.

The survey is concerned primarily with exercises or experiments in which a method of group working is adopted to replace or supplement an existing system. No attempt was made to apply close definitions to the term 'group working' since one objective of the survey was to help in the development of such a definition. The survey, however, deals primarily with deliberate attempts to reorganize work, and is therefore concerned with the creation of formal or orgainized work groups.

Nature and purpose of groups

Group size is specified in a majority of the cases, and ranges from 3 to 20 with an approximate mean of 9. Where reasons are given, the majority of exercises appear to have been undertaken in an attempt to overcome manifest problems, typically relating to productivity and quality. In certain cases changes external to the work-place prompted or provided the opportunity for the experiments, e.g. changes in output requirements, the introduction of mechanized equipment, the need for expansion, etc., whilst apparently in a minority of cases the exercise derived from some concern for the nature of the worker's role or tasks (e.g. to increase work variety, complaints about work conditions, etc.). Emphasis on the characteristics of the production system, e.g. throughput time and work-in-progress, is evident only in the case of group technology, whilst problems relating to worker absence and turnover are given as reasons for change in very few cases.

Work-group responsibilities*

The allocation of greater responsibility for the production process was a significant feature in many of the exercises. The responsibilities assigned to the work groups are listed in Exhibit 7.3. Two basic categories of group responsibility may be identified; firstly those responsibilities which call for *regular* decision making or actions from the group or its members, and secondly responsibilities which might not necessitate recurrent decision making or actions.

It is relevant to note that in respect of the fulfilment of responsibilities the exercises listed in Appendix C appear to fall into two categories. In a majority of cases groups have been established for the principal purpose of assuming greater responsibility for the day-to-day operation of their part of the production process. In such cases the groups have normally been given

---

* The distinction between responsibility and autonomy is made in Chapter 8.

Regular	Quality inspection
	Supply and requisition of materials
	Planning/scheduling of work
	Maintenance and housekeeping tasks
	Requisitioning of maintenance and repair
	Control of production in group
	Rectification work
	Work allocation⎱ or intermittent
	Rotation of jobs⎰
Intermittent	Problem solving
	Setting output goals
	Establishing break times
	Determining shift arrangements
	Appointment of representatives

Exhibit 7.3 Responsibilities of work groups

responsibility for tasks such as inspection, allocation of work, supply of materials, etc., i.e. regular 'day-to-day' production responsibilities. Such groups correspond to the autonomous work groups widely recognized in recent literature.

The second category includes the creation of those groups whose responsibilities are primarily of the intermittent, even perhaps 'once-only' type. Examples of the latter category include the following:

Exercise 33   Garment manufacturer, U.S.A.
    Groups of female workers participate in decision making in connection with job and associated piecework changes, and in the planning of such changes.

Exercise 37   Footwear manufacturer, Norway.
    Teams of production workers involved in group decision making with management in connection with the introduction of new models, i.e. allocation of models to groups, length of training period, division of labour and job assignments in groups.

Reference to the detailed descriptions of these latter exercises reveals their essentially *consultative* purpose. Such groups were not, it seems, established in order to permit the reorganization of planning and control activities previously undertaken by inspectors, supervisors, etc., but rather to facilitate communication, permit joint management/worker decision making, etc. The formation of such groups does not diminish the responsibilities of first-level supervision or of technical or service staff, whereas, in contrast, the formation of groups with 'regular' production responsibilities often appears to obviate the need for chargehands, separate quality inspectors, progress chasers, etc.

The distinction between these two categories of formal or organized work groups might be summarized as in Exhibit 7.4. We shall refer to the former as

Type of formal group	Primary purpose of group	Principal responsibilities might include:	Effects upon	
			Supervision	Specialist staff
Functional Work Group	Day-to-day operation of part of the production process	Quality inspection. Supply and requisition of materials. Planning/scheduling of work Maintenance and housekeeping tasks. Requisitioning of maintenance and repair. Production control and progressing. Work allocation. Work rotation. Rectification work	Work group assumes some of routine responsibilities of supervision. Supervision take on more of a technical or advisory role	Work group assumes some of routine responsibility of some ancilliary workers Group may be given direct access to specialist staffs
Consultative Group	Communication, representation and joint decision making with management and ancilliary workers	Appointment of representatives. Participation in problem solving. Setting goals for output and quality. Determining work-hours, breaks, shifts, etc.	Joint discussions, problem solving, decision making	Joint discussions, problem solving, decision making (usually with group/representative and supervisor)

Exhibit 7.4 Types of formal work group

Functional Work Groups and the latter as *Consultative Work Groups*. The groups as described above are of course stereotypes, and in practice it is likely that some overlap will exist, more so in the inclusion of the responsibilities of consultative groups in functional work groups than vice versa.

The greatest proportion (70 per cent) of exercises involved the establishment of 'team' and 'task' groups (see definitions earlier in chapter). In most cases the groups were fairly autonomous, working within quite broad terms of reference (team groups) or had some degree of control over the rate of work and the method adopted (task groups). The majority of task groups replaced groups where there was little opportunity for the use of discretion where work content and method were clearly specified (i.e. technological groups). Team groups, on the other hand, were created in similar proportions from both task and technological groups. Consultative groups were essentially created where the technology of the process controlled the worker's activities.

In general the pattern of development evident from these exercises involves the following movements:

(1) In cases where the main objectives of the exercise were quality or quantity improvement and cost reduction, the groups created were given greater flexibility over work method and over the rate of work (i.e. task groups). In over 60 per cent of the cases where assembly-line working was changed, task groups were established, whereas only one-third of changes introduced into other situations, e.g. job shops, resulted in the formation of this type of group.

(2) Apparently where companies were experiencing personnel problems and perhaps expressing concern about the motivational content of the work, they tended to provide much greater degrees of group autonomy (team groups).

(3) Task groups were typically established in assembly-line situations by a reduction in group size together with the addition of responsibility for inspection. In such cases work allocation usually remained the duty of the supervisor.

(4) In process operations changes often led to the formation of fairly autonomous or team groups. The changes were usually introduced as part of a productivity agreement.

It is of interest to note that whilst the exercises reported here were all selected because of their emphasis on formal group working, they demonstrate a surprisingly diverse pattern of changes. In particular we should note that whilst over one-third of the exercises involved the creation of virtually fully autonomous functional work groups ('team' groups) very many exercises involved the allocation of significantly less freedom to the work groups.

Results of group working

Improvement in quality is cited in a majority of those exercises in which results are discussed. A large proportion of exercises (40 per cent) also benefited from improvements in output. The dominance of such benefits corresponds in part to the reasons for the introduction of group working, and emphasizes the practical benefits of group working. Whilst also featuring amongst the benefits cited for the introduction of consultative groups, increased output and improved quality do not assume the same overriding importance for such groups. The other benefits associated with the introduction of consultative groups relate to improved worker attitudes, motivation and relationships.

References

1. Taylor, L. K., 'Worker participation in Sweden', *Industrial and Commercial Training* (January 1973), pp. 6–15.
2. Hamilton, A., 'Reaping the rewards of worker participation', *The Times* (6 November 1972), p. 20.
3. Cartwright, D., and Zander, A., *Group Dynamics—Research and Theory*, Harper and Row, 1960.
4. Lewin, K., *Resolving Social Conflicts*, Harper, 1948.
5. Schein, E. H., *Organisational Psychology*, Prentice-Hall, 1965.
6. Sayles, L., *Behaviour of Industrial Work Groups*, Wiley, 1958.
7. Dublin, R., *The World of Work*, Prentice-Hall, 1958.
8. Tannenbaum, A. S., *Control in Organisations*, McGraw-Hill, 1968.
9. Argyle, M., 'Group dynamics', *New Society* (2 November 1972), pp. 282–283.

CHAPTER 8

Functional Work Groups: Benefits and Basic Design Considerations

BENEFITS OF GROUP WORKING

Clearly, membership of some type of work group provides many potential benefits for the individual. Workers may benefit via increased confidence through recognition of important skills, through the development of social skills, through opportunity to exercise influence and assume a leadership role, whilst the group may provide an individual with support, encouragement, protection and security. Such benefits may derive from membership of various types of informal or formal groups. However, our concern is primarily with the creation of formal arrangements of workers, such groups being an inevitable feature of any production system. It is clear that factors associated with the technology and nature of the production process exert an important influence on the nature of functional work groups,[1,2] yet equally it is evident that within the constraints imposed by these factors some freedom exists in the 'design' of the work group, i.e. its purpose and organization. The survey of exercises in group working reported in the previous chapter and the results presented in Chapter 6 adequately demonstrate this latter point by illustrating the surprisingly large number of different approaches that have been adopted in the design of formal work groups. Our primary objective here, therefore, is to examine the benefits and some basic factors affecting the design of such formal *functional* work groups.

The desired attributes of work and of jobs have been examined in Chapter 5. It has been shown that in restructuring jobs the provision of meaningful and worthwhile work, work variety, perceived contribution to product utility, worker discretion, decision making, accountability, responsibility and autonomy, use of workers' skills, promotion prospects and social interaction are normally borne in mind. This list bears similarities with the descriptions developed by other authors, some of whom have approached the problem somewhat differently. Emery,[3] for example, in examining the characteristics of socio-technical systems identifies two primary requisites for the emergence of worker 'task orientation', i.e.

(1) The individual should have control over the materials and processes of the task.

(2) The structural characteristics of the task should be such as to induce forces on the individual toward aiding its completion or continuation.

Emery notes that in many situations these requirements are absent, and further that the demand for close coordination makes the delegation of responsibility for the task impossible. He argues that there is much greater scope in the development of group responsibility for group tasks, since if the individual's tasks are interdependent with the group task it is possible for the individual to be meaningfully related to his personal activity through his group task. It can be argued that a group task, because of its greater content and complexity, is more likely to provide a satisfactory basis for the allocation of responsibility and discretion than the smaller and simpler individual task. The larger group task is more likely to provide structural conditions conducive to goal setting and striving. If it has a measure of autonomy and a wide sharing of skills needed for its task, a group is also able to provide a degree of continuity in performance that is unlikely to be achieved by individuals under the control of a supervisor.[3]

Apart from the socio-psychological benefits of group membership, which may in any case derive from informal group membership, it can be argued that the existence of functional work groups facilitates the establishment of production targets and standards, provides greater opportunity for work variety and facilitates adaptation to change.

The determination of production targets and schedules, and the measurement of performance against such standards, is easier for larger production units than for individual tasks or for separate workers. The continued mechanization of production processes will add to their complexity thus making it more difficult to derive objectives, schedules, etc., appropriate for individual operators. Indeed, on occasions it may not be possible to establish clear demarcation between the responsibilities of individual workers.

Opportunity for increased work group variety for the individual members of a formal work group is provided through the possibility of job rotation. Thus the variety of work available to the individuals within such groups is likely to be greater than that which might be allocated to each individual. Equally, however, since job rotation within groups requires the agreement of those involved, the system may also accommodate workers who do not seek increased variety.

Disturbances caused through changes in product mix, design, output requirements, absenteeism, machine breakdown, etc., may necessitate the modification of production schedules and work allocations. Such decisions must often be made with minimum delay, yet depend upon close familiarity with the production system and necessitate flexibility of both workers and work system. Effective group working may provide such flexibility and may further permit such decisions to be taken within the group thus facilitating rapid response to external disturbances.

Clearly the required job attributes identified in Chapter 5 may be made available to some degree through the creation of formal functional work groups. Equally, however, it is clear that the provision of some such attributes may also be pursued by other means, e.g. job rotation, enlargement, etc. It seems therefore that in this respect the creation of formal functional groups is mainly beneficial in providing increased opportunity for the provision of these job and work attributes, whilst the extent to which they might be provided is dependent upon the *responsibilities* of the work group and their *autonomy* of operation.

In terms of job restructuring and work organization, formal functional group working therefore appears to constitute a major form of organizational change as identified in Chapter 5 and as such greatly facilitates work and job changes (see Exhibit 5.1). Furthermore, it will be seen that in the survey undertaken in Chapter 5 those exercises classified as 'self-organization' cover a large number of exercises in which formal function work groups were introduced.

BASIC DESIGN CONSIDERATIONS

Before examining work-group responsibility and autonomy—the key features of work-group design—we shall look at certain basic considerations which are likely to influence the design of functional work groups.

Group structure

In any group individuals are assigned a status or position, i.e. a collection of rights and duties. Each status position within the group has a role which the individual is expected to uphold; this is the dynamic aspect of his status. These status positions and roles—the basis of the group *structure*—along with other factors influence the behaviour of individuals and the performance of groups. Each individual possesses a set of group statuses as a result of membership of many groups, which results in conflicting pressure being brought to bear upon him. The greater the disparity between the values and norms of any two groups, the greater the conflict and the more difficult it will be for him to find a resolution of the opposing forces which are acting upon him. In many cases the formal organization will have within it an informal organization of different structure. This may lead to the individual being subject to formal rules which deviate from the informal, and thus he is faced with the choice between social unacceptability or managerial reprimand. The design of the formal structure should ensure that each position consists of a set of feasible functions, that members have the authority necessary to carry out tasks for which they are responsible and that there is effective communication of necessary information.

Group size

As groups increase in size, a smaller proportion of persons become central to the group, make decisions for it and communicate to the total membership. As an organization grows, a smaller proportion of members keep in touch with one another since there are too many persons for complete communication to occur, and many members may not even know one another. Communication tends to be impersonal in larger groups, a set of printed rules becoming essential for its functioning. Larger groups also tend to be less cohesive, whilst absenteeism,[4] output,[5] and the individual's contribution to decision making,[6,7] have also been shown to deteriorate with increasing group size.

Rice[8] has suggested that 'a group consisting of the smallest number that can perform a "whole" task and can satisfy the social and psychological needs of its members is alike from the point of view of task performance and of those performing it, the most satisfactory and effective group.' Miller and Rice[9] indicate that groups of different sizes have different characteristic patterns of behaviour: '. . . so far as we can tell at present, there are characteristic changes with each additional member up to five, six or even seven. Thereafter, as the group grows from 7 to between 11 and 16 members, though changes take place, the essential characteristics are those of the small face-to-face group. In such a group there are not so many members that they cannot sustain close and continuous personal relationships; but neither are there so few that the defection of one member can jeopardize the group's security.' They point out that larger groups may or may not be internally differentiated; undifferentiated large groups are usually very short-lived. Lacking the controls imposed by structured relationships at the work-group level, they are more prone to be dominated by irrational assumptions and beyond a size of twelve, or thereabouts, groups tend to split into subgroups.

Cohesion

The degree of cohesion in a group depends upon the attractiveness of the group to its members, the lower the attractiveness the less likely the members are to conform, and thus the lower the cohesion. An obvious influence on a group's cohesion and internal control is the degree of internal proximity between members. If members of a group sense a close interdependence among them for satisfactory completion of a task, they are constantly reminded of their closeness by the routine which they perform. If their combined efforts add up to a tangible result with which they can easily identify themselves, their sense of cohesion is magnified. A similar effect occurs if payment is related to group output and if the leader treats them as a group. Likert,[10] reviewing research in this area, concluded that work groups which have high peer-group loyalty and common goals appear to be effective in achieving their goals. If their goals are the achievement of high productivity and low waste, these are the goals they will accomplish. If, on the other hand, the character of their supervisor causes them to reject the objectives of the organization and set goals at variance with these objectives, the goals they establish can have strikingly adverse effects upon productivity.

Seashore,[11] examining the effects of cohesion in groups, found

(1) That highly cohesive groups showed much less within-group variation on productivity than low cohesive groups, but
(2) That the production variation between groups was greater for the more-cohesive than the less-cohesive groups.
(3) The high-cohesive groups were 'above' average in performance when they accepted company goals and below average when they rejected company goals, whereas the low-cohesive groups tended to be more average in performance.

If a group's members are so much alike as to be nearly indistinguishable, cohesion detracts from, rather than stimulates, productivity. In this extreme of cohesion, either the group does not possess a productive complement of skills or the group tends to focus so heavily on social activity that it neglects to attend to the demands of the environment.

The available research evidence suggests that group cohesion—a factor influencing labour stability and group performance—will be higher in groups whose members are similar in those respects which are important for interpersonal solidarity but complementarily different in terms of the skills required for coping with group tasks. Group payment systems may enhance the perceived degree of interdependence, and spatial proximity may also improve cohesion. Argyle,[12] similarly concludes that cohesiveness, important in respect of job satisfaction, low absenteeism, and turnover and productivity are developed under the following conditions: physical proximity and frequency of interaction, homogeneity of members, a group bonus or incentive system, small groups as well as a socially skilled leader and absence of disruptive personalities.

Cohesion clearly affects internal group relations and interaction, and through this affects social relations, satisfaction, withdrawal and conformity. An important disadvantage, however, derives from the possible development of animosity towards other groups. This results in reduced cooperation between groups and detraction from group objectives.

Goals and integration

Organizations are established for a specific purpose, performing work towards some end or goal. Organizational goals will be laid down or agreed for the work group; operatives' goals, however, designate the ends sought through actual operating policies of the organizations. The organization, in order to achieve its own goals, may subdivide the task and establish formal goals for the work group or the individual, concerning such factors as the level and quality of output produced. In order to achieve their goal, the work group in turn will establish its own operative goals. These will deal with areas such as output levels and earnings. The groups may also possess derived goals. Their influence will usually depend upon group cohesiveness, strong leadership and their power to disrupt the system.

Many factors influence whether or not groups will tend to fulfil both organizational and personal goals or only one of these. These factors have been divided into three classes, i.e. environmental factors such as cultural, social and technological climate in which the group exists; membership factors such as personal characteristics, backgrounds, relative statuses; dynamic factors such as organization, leadership, past performance, etc. Environmental factors will greatly influence group behaviour since they will determine possible interaction patterns. If groups are to be encouraged to fulfil organizational tasks the work environment must permit and, preferably, encourage the emergence of 'logical' groups. There are two possible approaches; firstly, formal groups can be established for the completion of a designated task; secondly, informal groups can be encouraged by the provision of opportunities for interaction, sufficient free time and favourable physical conditions. The degree to which logically designed groups come to serve psychological needs will also depend upon the managerial style and the degree of external control exercised.

Whether a group works effectively on an organizational task whilst at the same time providing satisfaction for its members depends in part on the group composition. For any group to work effectively there must be some degree of consensus on basic values and on a medium of communication. If personal backgrounds, values or status differentials prevent communication the group cannot perform well. An inadequate distribution of relevant skills and abilities may be another important membership problem. If the group fails in accomplishing its task because of resources and thereby develops a psychological sense of failure, it cannot develop the strength and cohesiveness to serve other psychological needs for its members.

The selection of members of a group aimed at bringing together individuals capable of harmonious interpersonal relationships was initially examined by Moreno[13] who described the process of classification as sociometry. Van Zelst[14] subsequently studied sociometrically selected work teams in the building industry and concluded that productivity and labour turnover were affected by such selection. The selection of teams was based upon individual preference of workmate. He also stressed the need for management to have a democratic approach to the government of workers as well as recognizing the importance of group relations and manifesting an interest in worker preferences.

Some production systems may necessitate a particular type of group arrangement. In other cases alternative socio-psychological systems may be possible within the constraints provided by the requirements of the processes and equipment. One objective of more recent research has been to identify the best fit between the technical and social systems and to introduce into any given situation the reforms needed to attain that fit. Miller and Rice[9] suggest that the conditions under which autonomous work groups are likely to be effective are:

(1) Closure or a sense of completion in finishing a meaningful unit of work.

(2) The group must be able to regulate its own activities and be judged by results. This necessitates a well-defined task boundary and established performance criteria. Not every member has to make all the decisions about his work in order to experience a feeling of autonomy or self-determination. If his own immediate group has some degree of decision making in which he has the opportunity to participate this may well satisfy his needs. Care must be taken not to make the group responsible for more than that with which it can cope at any point in time.

(3) The group size must be such that it can regulate its own activities and also provide satisfactory personal relationships. If the group is too large subgroups will form each possessing unique goals, possibly differing from each other; this will reduce cohesion in the main group and may lead to conflict.

(4) The group will be more stable if the range of skills required of group members is such that all members of the group can comprehend all the skills and could aspire to their acquisition. This facilitates communication between members and can aid group cohesiveness. The fewer the differences in prestige and status within a group, the more likely it is that the internal structure of a group will be stable and its members will accept internal leadership.

(5) The system should be flexible enough for the worker who becomes disaffected with one work group to have the opportunity to move to another engaged upon similar tasks.

Supervision

There is little doubt that the supervision of a work group can considerably affect the productivity of that group. As well as being affected by the supervisor's personality and the managerial style he adopts, the technical competence of the supervisor and his ability to organize and perform supervisory functions will influence many aspects of group development.

One early study of the influence of supervision upon the productivity of work groups was undertaken in an insurance company.[15] Here supervisors were moved between successful and unsuccessful groups and after a period of one year it was found that the previously unsuccessful groups under successful supervisors had significantly improved, whereas the opposite had occurred in the other situations. Supervisors' consideration of the needs of subordinates is often put forward as a reason for superior group performance. It has been concluded[16] that employee-oriented supervisors tend to obtain better output, motivation and worker satisfaction, and also that more successful supervisors are supportive, friendly and helpful rather than hostile, and also that they attempted to treat people sensitively and considerately.[17] Katz, Maccoby and Morse,[18] working in a life-insurance company also, concluded that 'employee-centred' supervisors tended to be in charge of high-producing groups, whilst in a similar type of study[19] it was concluded that men in high-producing groups more frequently described their supervisors as taking a personal interest in them, as being helpful in training them for better jobs and as being less punitive than supervisors in charge of low-producing sections.

It does appear, however, that different situations require different supervisory methods,[10] whilst the leader traits and methods which result in effective group performance are normally assumed to depend upon such situational variables as the group's objectives, its structure and the personalities of its members.

Participation in decision making with supervision in areas of direct group interest is thought by some researchers to result in both greater job satisfaction and increased productivity. An experiment at the Harwood Manufacturing Plant[20] showed that productivity increased when work groups participated in setting production goals, whilst it has been indicated[21] that groups who set their own production goals showed a significantly greater increase in production than those not setting their own goals. Similarly, dramatic output increase had been reported[22] when a group of workers at a toy factory were given control over the speed of group working.

Change and work groups   ·

Strong peer groups at the rank-and-file level may be resistant to essential organizational change.[11] Resistance to change may be manifest through the withdrawal of cooperation, aggression and grievance action, or may result in counter-productive behaviour such as low efficiency, output restriction, high absenteeism and perhaps labour turnover.

Efforts to change behaviour can be resisted by pressures from within the group. The same range of influence can also be used as a medium of change. The group may itself become the target of change for its members, e.g. its leadership, norms, standards, etc. and finally the group may be used as an agent of change. The latter requires some feelings of common purpose between those who are to be affected by change and those seeking it. The more attractive the group to its members the greater its influence. If the attitudes, values, etc. which are to be changed are held to be important to the group, the influence of the group will be greater. Members with greater prestige in the group will be able to exert greater influence, whilst attempts to change individuals or parts of the group which might lead to the breaking of the accepted rules or standards of the group will also meet resistance.

It follows that for many kinds of organizational change to be successful it is necessary to deal with groups as targets of change. Strong pressures supporting change can be created by establishing shared perception by members of the need for change, thus placing the source of pressure within the group.

The benefits of group decision making as a means of changing group behaviour have been appreciated for some considerable time. Lewin[23] demonstrated the importance of such methods in changing habits whilst Bavelas[24] reported the permanency of change following the group decisions of a team of sewing-machine operators. Coch and French[25] described the favourable outcomes resulting from the involvement of groups in planning changes at Harwood Manufacturing Company, whilst later French *et al.*[26] studied the effects of group participation in a Norwegian factory and pointed out the need for the participation in decision making to be relevant to the task at hand.

The participation of the group in the decision-making process generally requires the adoption of a new management style. Likert[10] described a 'participative' style of management which investigation has shown to be a more appropriate style for some modern industrial organizations than more autocratic styles, particularly where rapid changes in technology are taking place.

Comparatively little has been written on the introduction of planned change on a collaborative basis,[27] however the process has recently attracted renewed attention in Europe, largely as a result of industrial experiments deriving from work in the field of industrial democracy in Norway.

GROUP RESPONSIBILITIES AND AUTONOMY

Responsibilities allocated to functional work groups have been examined earlier. The concept of group autonomy has also been introduced, and it has been demonstrated that the degree of group autonomy and its responsibilities influences the manner and extent of the provision of certain desirable job attributes. The group's formal responsibilities comprise those tasks and duties for which they are individually or collectively accountable to their superiors.*

Autonomy is associated with freedom from controls and the facility for self-organization and regulation. The autonomy and responsibilities of work groups are closely related, and in practice probably positively correlated. Although many authors avoid the distinction, the use of the two terms facilitates the description of methods of group organization and working. It seems unlikely that we would reach any consensus in any attempt to devise rules relating to the degree of autonomy, or the extent of the responsibilities to be allocated to work groups, since in both cases the matter of degree is a function of circumstance, objectives, etc. It will be useful, however, to identify the dimensions of autonomy in group working, and to list the responsibilities that might be assumed by such groups. This will permit us not only to develop an idea of the nature of the fully autonomous and responsible work group but will also provide checklists for use in the design of work groups and against which to measure such groups.

Davis,[28] having studied several exercises in job design/redesign, concluded that where small organizational units, or work groups, are required, group structures having the following features appeared to lead to improved performance:

(a) Group composition that permits self-regulation of the group's functioning.

---

* Regular discharge of these responsibilities will normally be recognized by the provision of rewards, i.e. wages, whilst discharge of informal responsibilities normally associated with informal organization is not normally recognized in this manner.

Dimension of Autonomy	High Autonomy	Examples/Means $\longrightarrow$	Low Autonomy
(1) Goals			
(a) Qualitative: what to produce	Selection of product	$\longrightarrow$	Involvement in product design
			Allocation of new products
(b) Quantitative: how much to produce	Determination of output targets and goals	$\longrightarrow$	Scheduling within targets
			Set subgoals
terms of payment and other sanctions	Specification	$\longrightarrow$	Negotiation with management
(2) Performance			
(a) Decide where to work	Location of plant	$\longrightarrow$	Location of group in department
			Location of workers within group
(b) Decide when to work: determine total working hours	Specification	$\longrightarrow$	Negotiation with management
determine arrangement of hours in week	Total freedom	$\longrightarrow$	'Flextime'
determine if workers may leave work during normal hours	Total freedom	$\longrightarrow$	Agree with supervision
determine and assign overtime	Extent/need for overtime	$\longrightarrow$	Satisfaction of given overtime requirement
fix breaks etc.	Total freedom	$\longrightarrow$	Within given period
(c) Decide to engage in other activities	As required	$\longrightarrow$	When production targets satisfied

99

	Design methods	
(3) Production Method (Group or individual decisions)		→ Method improvement / Participation in job design
(4) Distribution of tasks	Initial specification, allocation and rotation	→ Rotation only
(5) Group Members		
(a) Select and appoint new members	Group decision	→ Group/management decision
(b) Expel unwanted members	Group decision	→ Request management action
(c) Discipline members	Group decision and action	→ Request management action
(d) Train new members	Induction, basic, speed and flexibility training	→ Aspect of training
(6) Leadership		
(a) 'Internal' leader: determine whether required	Determine role and responsibility	→ Against given role
selection if required	Select	→ Consult with management
(b) 'External' leader: determine whether required	Determine role and responsibility	→ Against given role
selection, if required	Select	→ Consult with management

Exhibit 8.1 Dimensions of work-group autonomy

(b) Group composition that deliberately provides for the full range of skills required to carry out all the tasks required in an activity cycle.

(c) Delegation of authority, formal or informal, to the group for self-assignment of tasks and roles to group members.

(d) Group structure that permits internal communication.

(e) A reward system for joint output.

In contrast, Gulowsen[29] developing criteria for the measurement of work-group autonomy specifies a more extensive list, embracing far greater degrees of autonomy than implied by Davis. Whilst Davis' list probably reflects more accurately the likely degree of autonomy of formal functional work groups, a treatment based on Gulowsen is used here since our objective is comprehensiveness. The list given in Exhibit 8.1 is modified slightly from the original in order that we might distinguish satisfactorily between group autonomy and group responsibilities (the latter was not dealt with by Gulowsen). These dimensions of autonomy relate to the freedom of the group to manage their affairs. They are *organizational dimensions*, unlike the responsibilities listed in Exhibit 8.2, which are related to tasks, i.e. the dimensions against which production performance is measured.

---

(1) *Materials and Products*
Inspection of finished items
Inspection of materials and components
Fault finding in products
Rectification/repair of defects
Order and/or collect own materials

(2) *Equipment*
Set up equipment and adjust
Request/schedule/undertake maintenance work
Request/undertake repair work
Request/collect tools

(3) *Work Area*
Cleanliness of work-place/area
Safety at work-place/in area

(4) *Communications*
Report/record output
Report/record defectives/quality performance
Report/record work in progress/material stocks
Progress items through preceding operation/group operations

Exhibit 8.2 Responsibilities of work group

In practice, the autonomy of few groups will extend to the control of the formulation of goals (1). Some degree of involvement in goal formulation has been given as a desirable job attribute (Chapter 5); however, in practice this is unlikely to embrace the selection of the product to be made, the determination of overall output targets or complete freedom in the specification of rewards.

Nor is it likely that production groups will exercise much influence on the location of the work place (2a) or the determination of total work hours (2b). However, certain consultative groups (Chapter 7) may well have been established with the purpose of contribution to such decisions or negotiating terms on behalf of their members. Autonomy in respect of choice of production method (3) and internal distribution of tasks (4) is relevant only if real alternatives exist. For example, in process industries production methods may be determined by the technology employed, whilst heterogeneous tasks and varied skills may prohibit alternative task allocations.

These dimensions of autonomy together with the responsibilities listed in Exhibit 8.2 and the other basic design considerations identified in this chapter will be employed in the examination of group working in mass production in the following chapter.

References

1. Sayles, L., *Behaviour and Industrial Work Groups*, Wiley, 1958.

2. Taylor, J. C., 'Some effects of technology in organisational change', *Human Relations*, **24** (1971), pp. 105–123.

3. Emery, F. R., 'Characteristics of socio-technical systems', in Davis and Taylor (Eds.), *Design of Jobs*, Penguin, 1972.

4. Mayo, E., and Lombard, G. F. F., *Teamwork and Labour Turnover in Aircraft Industry in Southern California*, Boston, Harvard University, 1944.

5. Marriott, R., 'Size of working group and output', *Occupational Psychology*, **23** (1949), pp. 47–57.

6. Hare, A. P., 'Interaction and consensus in different sized groups', *American Sociological Review*, **17** (1952), pp. 261–267.

7. Slater, P. E., 'Contrasting correlates of group size', *Sociometry*, **21** (1958), pp. 129–139.

8. Rice, A. K., *Productivity and Social Organisation*, Tavistock, 1967.

9. Miller, E. J., and Rice, A. K., *Systems of Organisation*, Tavistock, 1967.

10. Likert, R., *New Patterns of Management*, McGraw-Hill, 1961.

11. Seashore, S., *Group Cohesion in the Industrial Work Group*, University of Michigan, Survey Research Centre, 1954.

12. Argyle, M., *The Social Psychology of Work*, Allen Lane, 1972.

13. Moreno, J. L., *Who Shall Survive?*, Nervous and Mental Disease Publishing Company, 1937.

14. Van Zelst, R. H., 'Sociometrically selected work teams', *Personnel Psychology* (1952), pp. 175–186.

15. Feldman, H., *Problems in Labour Relations*, Macmillan, 1937.

16. Davis, K., *Human Relations at Work*, McGraw-Hill, 1962.

17. Likert, R., 'A motivational approach to a modified theory of organisation and management', in Hare (Ed.), *Modern Organisational Theory*, Wiley, 1959.

18. Katz, D., Maccoby, N., and Morse, N. C., *Productivity, Supervision and Morale in an Office Situation*, University of Michigan Institute for Social Research, 1950.

19. Katz, D., Maccoby, N., Gurin, G., and Floor, L. G., *Productivity, Supervision and Morale among Railroad Workers*, University of Michigan Institute for Social Research, 1951.

20. French, J. P. R., 'Field experiments—changing group productivity', in Miller (Ed.), *Experiments in Social Progress*, McGraw-Hill, 1950.

21. Lawrence, L. C., and Smith, P. C., 'Group decision and employee participation', *Journal of Applied Psychology*, **39** (1955), pp. 334–337.

22. Whyte, W. F., *Money and Motivation*, Harper, 1955.

23. Lewin, K., 'Frontiers in group dynamics', *Human Relations*, **1** (1947), pp. 5–41.

24. Maier, N. R. F., *Psychology in Industry*, Houghton Mifflin, 1946.

25. Coch, L., and French, J. P. R., 'Overcoming resistance to change', *Human Relations*, **4**, 1 (1948), pp. 512–533.

26. French, J. P. R., Israel, J., and Aas, J., 'An experiment on participation in a Norwegian factory', *Human Relations*, **13** (1960), pp. 3–19.

27. Trist, E., 'The professional facilitation of planned change in organisations, *XVI International Conference on Applied Psychology 1968*, Swets and Zeitlinger, 1968.

28. Davis, L. E., 'The design of jobs', *Industrial Relations*, **6** (1966), pp. 21–45.

29. Gulowsen, J., 'A measure of work group autonomy', in Davis and Taylor (Eds.), *Design of Jobs*, Penguin, 1972.

Part 5
Group Working and Mass Production

The scope and opportunities for the formation of formal functional work groups in flow-line and other mass-production systems.

CHAPTER 9

Formal Work Groups and Production Flow Lines

It has been argued, and hopefully demonstrated, that the concept of 'group working' is of considerable importance in connection with the development of mass production, and in particular with reference to restructuring jobs and work organization. The previous two chapters have been devoted to a closer study of this concept, during which it was noted that group working could not be considered as a system of production, since all such systems in practice involved the use of groups of workers.

Closer examination of the nature of group working revealed that the changes discussed in Chapter 6, in which group working had been introduced, were primarily examples of organizational change rather than production-system changes. Group working was subsequently associated with the creation of formal functional and consultative work groups (Chapter 7) the former being seen as a means for providing the organizational change from which certain types of job restructuring might derive (Chapter 8).

The benefits gained from the creation of formal functional work groups were found to be associated with the extent of the responsibilities of the work group and their autonomy of operation (Chapter 8). Hence in this and the following chapter we shall look not only at the manner in which formal functional work groups might be created within the constraints imposed by the various types of mass production, but also at the degree to which such groups might assume both responsibilities and autonomy. The checklists presented in Chapter 8 will be employed for the assessment of functional work groups, i.e.

(A) *Responsibilities* relating to
    (1) Materials and products.
    (2) Equipment.
    (3) Work area.
    (4) Communications.
(B) *Autonomy* in respect of
    (1) Goals.
    (2) Performance.
    (3) Production method.
    (4) Distribution of tasks.

(5) Group members.
(6) Leadership.

The functional work groups described in this chapter will also be assessed against certain of the basic design considerations identified in the previous chapter, i.e.

(C) *Basic design considerations* relating to
   (1) Group structure.
   (2) Group size.
   (3) Group cohesion.
   (4) Group goals and integration.

GROUP WORKING AND MANUAL FLOW LINES

Two forms of pacing may be evident at work stations on manual flow lines. Non-mechanical lines may subject workers to a form of *operator pacing*. Such lines generally function with buffer stocks between stations. Although such stocks permit short-term differences to exist between the times required for operations at stations, the average time for operations must be equal to avoid either the congestion of work on the line or delays at stations. This requirement gives rise to a form of operator- or self-pacing at stations, the smaller the buffer stocks between stations the higher the pacing effect.

Moving-belt lines provide a mechanical pacing effect since excessive operation time at stations may mean that an item passes from the station incomplete (fixed-item lines) or a subsequent item passes by the station unattended (removable-item lines). Other things being equal, the pacing effect is reduced by increasing station length and reducing line speed together with the closer spacing of items on the line.

In both cases, therefore, the extent of either mechanical or operator pacing can be manipulated by variation of certain line parameters. It is this pacing effect (either mechanical or operator), together with the length of the operations at the stations (i.e. the cycle time) and the number of stations on the line, which largely determines the extent to which workers on such lines might assume some of the characteristics of autonomous responsible groups.

(1) HIGHLY RATIONALIZED AND PACED LINES

Basic design considerations in group working

Workers on such lines are normally unable to stop work on their own operation for more than a short period of time, unless replaced by another worker, thus the degree of interaction during the work period is limited, even when some station overlap exists or workers work together at stations. Workers on fixed-item, moving-belt lines may move out of their stations, perhaps for considerable distances up or down the line, especially when there are more items than workers on the line; however, whilst this may permit workers to gather together at parts of the line, their interaction is severely restricted. Although these workers may be in close proximity, and whilst their number may correspond closely to a desired group size, little group structure or cohesion may exist. Certainly such workers are interdependent, they may be paid on a group wage system, have common skills and constitute a logical grouping, nevertheless they clearly differ fundamentally from the type of functional group identified previously. It is unlikely that members of the group will feel much sense of completion although the group may together assemble a complete item, nor are they able, as a group, to regulate their own activities even though there are well-defined task boundaries and clear criteria for measuring performance. Communication between the members of the group might be limited by the physical arrangement of the stations, and communication with supervision, etc., is also likely to be limited to defined channels and procedures.

Group autonomy and responsibility

Some degree of autonomy in the distribution of tasks may be available through job rotation; however, the absence of a clear group structure and interpersonal contact may limit the opportunities for self-organized rotation. Alternatively, rotation may be organized for the group to provide job enlargement. The difficulty of organizing such changes within the group limits the degree to which the group may influence their place of work, and the place of the individuals within the work area. The short cycle time and the close interdependence of workers ensures that all workers must be present at their work-places at the same time, thus there is little opportunity for group determination of working hours, breaks, etc., unless all members of the group act in concert.

The emergence of a leader within the group will in general depend upon a group consensus which may be prevented by the absence of structure and cohesion. Similarly, formal contact with other staff will generally take place through formal representatives. Group autonomy in the selection and training of members will also be severely limited, although new workers may well be trained by workers on the line—this will generally be organized by supervision.

The responsibilities of members of the group might cover some of the items listed in Exhibit 8.2; however, in practice such responsibilities will generally have been allocated to workers as part of their regular work in order to maintain line balance, e.g. inspection by workers at the end of the line. Furthermore, tasks which occur on an irregular basis, such as equipment maintenance, replenishment of stocks, etc., will not usually be undertaken as this would disrupt the work balance and the flow of work. Generally tasks will be undertaken by ancillary workers whilst others will be responsible for temporarily taking on the jobs of absentees. Hence the design of the system accommodates the need to perform intermittent duties, as well as providing a means for dealing with internal disturbances, thus the scope for internal group organization is minimized.

(2) LINES WITH 'INCREASED' CYCLE TIME AND 'REDUCED' PACING

Reduced pacing and increased cycle time permits greater freedom for workers, thus lines with say up to twelve to fifteeen workers may offer conditions more appropriate to the existence of formal functional work groups. Furthermore, it may be possible to arrange comparatively short lines in such a manner that interpersonal contact is facilitated (e.g. square or circular, rather than linear configurations). This proximity, together with the reduction of the pacing effect and the possibility of reduction of the dependence of the group on outside support to maintain regular work flow, may give rise to improved group identity, integration and cohesion. Thus self-organized job rotation becomes viable together with the assumption of responsibility for indirect and intermittent tasks previously prohibited through the need for continuous manning of stations. The greater cycle time enables underloaded, or particularly proficient, workers to build up sufficient 'slack time' for them to move elsewhere on the line or perform indirect tasks. The longer operations facilitate the redistribution of tasks by workers either to improve work balance, cover for absentees, compensate for the introduction of a trainee or failure of equipment, accommodate model changes, etc. The smaller group size may help individuals to identify themselves more closely with the group task and should permit the development of the consensus necessary for much group decision making. For example, agreement may be reached on the selection and role of an informal leader or representative, the manipulation of work hours, timing of breaks, etc. The longer cycle time, and the freedom of workers to leave their stations, may permit the internal allocation, sharing or alternation of responsibilities including inspection of items and material, supply of parts, rectification and completion of paper work.

If sufficiently independent of subsequent stages in the production process, absenteeism in the group, or other short-term changes in group size, whilst necessarily leading to changes in output levels may be accommodated within the group through work reallocation. Such conditions may also facilitate short-term variations of schedules and output targets.

Workers in these groups may even choose a work allocation which leads in effect to the abandonment of flow-line working. They may choose to work together or in smaller teams, either collectively (see next chapter) or as shorter lines. Alternatively, individual assembly may be employed. In theory some of these alternatives may be possible on highly rationalized, closely paced lines. For example, individual assembly can be provided whilst using a moving-belt, fixed-item line for transport, as in gearbox assembly on the carousel conveyor in Case C. However, other conditions on such lines may prevent such methods being introduced as a result of group initiative. In practice, therefore, the development of structured integrated and cohesive work groups able to exercise group autonomy is likely to be influenced mainly by the degree of work pacing, the extent of work rationalization and the number of workers on lines. In general their creation may be easier on non-mechanical lines since the extent to which the pacing effect can be reduced on moving-belt lines may be limited by mechanical considerations and furthermore the use of such a mechanical handling device will probably limit the extent to which work may be reallocated within the group.

Operating characteristics

Exhibit 9.1 shows for a non-mechanical line the relationship between line length, interstation buffer capacity and line output efficiency.* The graph,

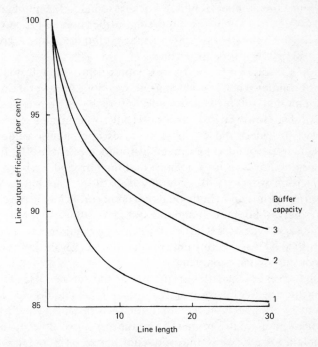

Exhibit 9.1 Operating characteristics of non-mechanical
flow lines

which describes a notionally balanced line (e.g. same operation-time distributions at stations), indicates that for any line length improved line efficiency is obtained with larger buffer stocks. Also, for given buffer capacity, shorter lines are more efficient, as fewer stations are dependent upon the supply of items from preceding stations. Since, for a given product work content, shorter lines

---

* The curves were obtained by digital-computer simulation and describe lines with all stations having normally distributed work times with a mean of 10 units and coefficient of variation of 0·3.

necessitate a longer cycle time, these requirements for line efficiency are consistent with the requirements for effective group working.

Much the same argument applies to moving-belt type lines except that in this case the situation is complicated slightly by the fact that line efficiency is influenced not only by the idle time occurring at stations but also by the production of incomplete items. It can again be shown that, other things being equal, idle time and the proportion of incomplete items produced improve with reducing line length and reducing mechanical pacing.

A longer cycle time also facilitates line balancing. The replacement of a single line by two or more shorter lines introduces some degree of freedom in the choice of cycle times, hence some discretion in the determination of group size may exist.

The introduction of shorter lines provides an opportunity to reduce work in progress. Since the number of buffer stocks is equal to the number of stations less one, fewer stocks will be required. Furthermore, as demonstrated in Exhibit 9.1, the interstation buffer capacity required for a given output efficiency will be lower for shorter lines.

Several of the cases referred to in Chapter 6 provide information on the problems of training when introducing enlarged jobs. In several cases workers were given periods of further training before being transferred to longer cycle-time jobs. However, it is also noticeable that in many cases no extra training was provided, indeed in some cases apprehension about training requirements were found to have been ill-founded (e.g. Case J). It is perhaps more important, however, to recognize that the introduction of some form of functional group working will generally lead to the enrichment of the job, which in turn depends upon individuals becoming proficient in the execution of a greater variety of both direct and indirect tasks—thus the ability to perform more of the same type of task is a minor training requirement, and one which may be justified by the resulting improvement in worker flexibility, seem to be advantageous in so many companies.

Additional costs associated with the use of shorter, less paced lines, which may also include higher investment in tools, equipment and space, are offset to some extent by the increased flexibility of such systems. The flexibility and internal organization of the group reduces the disruptive effect of absenteeism, equipment failures, etc., whilst the production of several models is more readily accomplished on several short lines; thus the need to rebalance and set-up lines for model changes is reduced. Output changes are more readily achieved by the addition or subtraction of one or more smaller lines, and the use of larger cycle times helps reduce the amount of non-productive work (e.g. handling and transport) required.

Examples

The examples below and in the following pages derive from the cases listed in Exhibit 6.1 and from other published work. Example 2 describes a situation in which lines were changed in the manner discussed above, whilst the remaining examples describe newly formed groups.

Example 1: tape-recorder assembly (Case 13—Exhibit 6.1)

A non-mechanical flow-line system is employed in the assembly of the main part of the record/playback head for tape recorders. A group of nine girls are together responsible for the assembly of parts for eleven types of head. Cycle time is 2·35 minutes, whilst buffer stocks consisting of several jigs, each containing twenty items, exist between stations. Assembly and solder workers are free to change jobs, indeed some such movement is necessary to maintain line balance. Workers are responsible for replenishing their supply of parts and materials, whilst job rotation is voluntary and self-organized.

'Flextime' working has been introduced on an experimental basis for thirty-five women workers employed on two non-mechanical tape-recorder final assembly lines. Each worker on the line may choose her own hours daily within certain constraints, the possibility of such individual flexibility being provided by substantial buffer stocks between stations.

Example 2: electrical-power-supply equipment[1]

Assembly work in electrical engineering, previously undertaken by males employed on a highly rationalized flow line, was modified to provide groups of three to five men; the group worked in effect as short, large-cycle-time, and low-paced, non-mechanical lines. They were responsible for organizing their own work, allocating tasks and conducting their own quality audit.

Example 3: wheel/tyre pre-assembly (Case A2—Appendix B)

The wheel/tyre pre-assembly team consists of ten workers, one of whom is selected by the team as a spokesman. The spokesman also plans the work for the group against a weekly output schedule. The group delivers wheels to each of four truck-assembly lines as well as to a spares department. Three of the workers in the group work on a rationalized flow line. These operatives move to less-paced jobs at regular intervals, otherwise job rotation within the group is self-organized. Absenteeism in the group is either covered by the team, or alternatively the spokesman requests extra labour. The group holds monthly meetings with their foreman and the engineer responsible for the area, and no quality control is applied to the group, items being inspected only when assembled to completed vehicles elsewhere in the plant. The group is paid on a group bonus system.

Example 4: motor-car-engine assembly (Case B—Appendix B)

Seven assembly groups of three workers are located alongside an automatic conveyor. Apart from some pre-assembly work, finished engines are completely assembled in each group, the total work content of this final assembly being thirty minutes. Each group may choose to divide the work between them, each working on average for ten minutes on each engine, or each member may assemble a complete engine. Engines are moved by hand in the group, but transported to and from the group mechanically.

(3) DIVIDED AND STATION-GROUPED LINES

The studies reported in Chapter 6 identified several examples of the structural modification of lines to provide improved conditions for group working. Such methods for modifying non-mechanical flow lines are discussed below.

Divided lines

The division of a long line into a number of smaller subunits without changing total line length, cycle time or the degree of pacing, may lead to the creation of improved group identity. In some cases the nature of the product and the manner of assembly might provide a type of 'natural break' in lines; alternatively, it may be possible to design a line to facilitate the creation of smaller groups, perhaps divided by large buffer stocks. This division of lines may overcome two of the problems associated with long, closely paced lines, by providing for smaller groups of workers in closer proximity. These smaller and closer groups may acquire some degree of cohesion and integration that would otherwise have been impossible. They may assume autonomy, covering for example the allocation and rotation of tasks, selection of a leader, and providing adequate 'buffers' exist between groups they may determine their own hours, breaks, etc. Inspection between groups may be convenient if responsibility for repair and rectification is given to the group; however, such additional irregular work can only be undertaken in conditions of low pacing.

The reduction of pacing and increased cycle time brings precisely the same advantages as identified previously, paramount amongst which is some freedom for workers to leave their stations to help colleagues, perform other duties, etc. Such conditions enable these groups to assume the autonomy and responsibilities discussed previously. However, even without this freedom the division of lines appears to offer some benefits.

Operating characteristics of divided lines

The introduction of large buffer stocks between stations, for the purposes of line division, may lead to an improvement in line efficiency since the size of the buffer between any two stations will influence idle time at these stations (although in practice, beyond a certain level, buffer capacity provides only minimal improvement—Exhibit 9.1). More importantly, however, sufficiently large buffers at certain parts of the line may, if utilized, in effect divide the line into separate lines, and since shorter lines are more efficient, even with the same cycle time and interstation buffer (see Exhibit 9.1), divided lines may incur less idle time. No operating benefit results for either fixed- or removable-item, moving-belt lines, unless workers at stations after the dividing buffer stocks are required to place items on the line, in which case a reduction in idle time may be obtained at these stations. The proportion of rejects produced is unaffected unless inspection takes place between the sections of the divided line. In neither case, however, is line performance adversely affected, although the introduction of buffer stocks complicates handling requirements and adds to space and work-in-progress costs. To offset this these stocks provide some degree of protection against failures of equipment at stations and of the moving belt itself.

Examples

The examples below, indicate the extent to which line division may promote functional group working. Examples 2 and 3 describe both modified and newly designed lines.

Example 1: electricity-meter assembly (Case 18—Exhibit 6.1)

Assembly of domestic electricity meters is undertaken by thirty-eight operatives (mainly females) on a non-mechanical assembly line. The line is divided into three groups, each concerned with the assembly of a major part, or the final assembly of all parts of the meter. Buffer stocks of between ten and thirty items exist between stations, whilst the groups are divided by larger stocks. Because of absenteeism on the line (approximately 10 per cent), and in order to accommodate the assembly of batches of meters of different designs, frequent deployment of labour is necessary within groups. Workers within the groups were often required to move to different operations or divide their time between two or more operations, such deployment being organized by the chargehands responsible for each group.

Example 2: assembly in an electronics company[2]

In order to attempt to reduce inefficiencies associated with delays caused by model changes, shortages of materials, turnover and absenteeism, a 104-station flow line manned by female workers was divided into five more or less equal groups of workers. These groups were separated by buffers of items containing approximately one hour's work. The groups were associated with a clearly defined set of tasks, and were positioned in such a way as to improve interpersonal 'contact'. Each group was given its own inspector, so that quicker reporting of faults and feedback was possible. Each group was also responsible to its own supervisor.

No change was made on the work-cycle time; however, improvements were obtained in respect of waiting time, absenteeism and quality.

Example 3: truck assembly (Case A3—Appendix B)

The truck-assembly method considered by this company in 1971 provides an example of a divided-line approach to flow-line working. An average cycle time of twenty-five minutes was to be used in the assembly of several models of truck. The complete line was to consist of four three-station lines divided by buffer stocks. Stations on each of these parts of the divided line were to be manned by approximately six workers. An indexing system was to be employed, the index movement being controlled by signals from the stations on the line.

Station-grouped lines

In certain cases it may be possible to deploy work groups at the stations of a line. Such an arrangement may be employed when items are sufficiently large to permit two or more people to work together, or alternatively workers may operate in parallel, each employed on one item. The former case, which is more likely to be appropriate for fixed-item, moving-belt lines, might in fact provide a series of perhaps collective working groups of the type discussed in the next chapter. Workers in station groups on non-mechanical lines might also work together on a single item, but work groupings at stations on removable-item, moving-belt lines will almost certainly require workers to be employed on separate items. In this latter case each station is in effect paralleled, and whilst cooperation between workers may be limited, each worker will have a cycle time greater than would have existed had all workers been deployed at separate sequential stations.

Station grouping, even in the parallel-station situation, may provide some of the benefits of divided lines by providing for small groups of people in close proximity. Whilst the degree of autonomy will be severely limited, their proximity may facilitate some social interaction and perhaps facilitate consensus and communication. Some degree of task allocation may be undertaken by the group to accommodate differing performance levels of members of the group. Rotation between different work areas, but not different operations, will be possible; sufficiently large buffers between grouped or paralleled stations may permit greater individual flexibility in respect of break times, etc. since in sufficiently large groups the short-term absence of one person may be covered by the remaining workers, and will in any case have a lesser effect on the following buffer stock than if that station had been manned by a single operator. In fact, on such lines individual flexibility of hours may be possible provided stations are paralleled or workers are able to perform the jobs of temporary absentees.

Such groups may undertake their own training providing the required average station performance is maintained, whilst sufficient flexibility to permit individual variation of break times and hours should also enable workers in groups to perform intermittent tasks such as rectification work, materials supply, etc.

Operating characteristics of station-grouped lines

The paralleling of stations on any type of line is likely to facilitate line balancing, since the provision of a good balance depends, amongst other things, upon the cycle time available at a station.[3] Paralleling of stations on non-mechanical lines improves line performance, since, for a given buffer stock between stations, less idle time is incurred at stations. This is illustrated by Exhibit 9.2 which compares a thirty-operator line in which there are two operators at each station, each performing the same operations, with two single operator/station, 15-station lines.* In both cases lines are notionally balanced. Average idle time for each operator is less for the paralleled line for all total buffer stock conditions. Alternatively, for a given line operating efficiency the total buffer capacity provided for the paralleled line is less than that of the equivalent single operator/station lines.

* Curves were obtained using digital-computer simulation of non-mechanical lines with each operator having a normally distributed work time with a mean of 10 units and coefficient of variation of 0·2.

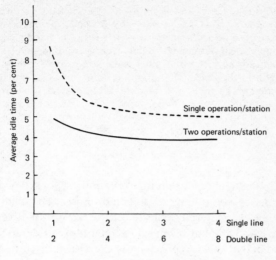

Exhibit 9.2 Buffer capacity

Station grouping in which workers work together on a single item, or indeed any similar grouping arrangement in which there are fewer items than workers on the line, provides the benefit of reduced work-in-progress and space requirements in comparison to a single operator/station/item line.

Examples

Two examples of functional-work-group formation through station grouping are given below. Example 1 shows how groups may be employed at stations on a moving-belt line whilst the second example illustrates the use of the station-grouping principle in a far more radical manner.

Example 1: truck assembly (Case A2—Appendix B)

Eight groups, each of four to six workers, operate at adjacent sections of the moving-belt, fixed-item truck-assembly line. Cycle time on the line is twenty-five minutes and group autonomy covers allocation of work, job rotation and quality responsibility. The workers in each group may decide to work together or on separate tasks and are free to rotate as required. Each group has one truck chassis to work on at their station. Responsibility for quality derives simply from the fact that faults are easily traceable to work teams. No representatives are elected.

Example 2: car assembly (Case A1—Appendix B)

This car plant is designed to provide two basic systems for vehicle-body assembly. A moving-belt, fixed-item line system can be employed, whilst an alternative 'dock' system is also available, which provides almost total freedom from mechanical pacing. Since for complete assembly a vehicle must pass sequentially through areas in which one of these systems is employed, the dock system described below is in fact a grouped station on a longer line.

The 'dock' system. The fifteen to twenty-five workers in a dock area are divided into groups of two to three workers, each responsible for undertaking the same operations on the vehicle; thus each dock area consists of a series of paralleled identical work stations. The group at each station may obtain vehicles from the preceding buffer stock and discharge them to the succeeding stock area. The body remains stationary whilst being worked upon in the dock. The group may allocate work amongst themselves, rotate jobs and determine their own breaks—being protected by large stocks. The paralleling of the stations also facilitates freedom for the groups since, providing all groups do not choose to stop work at once, work is still undertaken in the area, although at a lower rate.

DISCUSSION

Ostensibly it might appear that the creation of formal work groups in flow-line systems is something of a compromise. The characteristics of flow-line working are generally considered to prevent or severely limit the provision of those job attributes which are generally considered to be desirable (Chapter 5) and in particular to prevent the exercise of discretion by workers.

We have seen, both from the review of the published accounts of group-working exercises and from the case studies that very many functional work groups have been created in such flow-line systems. Equally, it is clear that such arrangements are virtually impossible unless some degree of worker freedom is available, such freedom being influenced primarily by the cycle time of operations and in the degree of work-pacing.

Several examples were provided of the creation of group working on lines where cycle time had been increased, the number of stations reduced and the pacing effect—normally operator pacing—had been reduced. This approach to the provision of group working was emphasized in this chapter, since it is likely that for most companies this 'solution' may be more appropriate than either line division or station grouping. Equally, however, we should note that some of the best examples of autonomous responsible work groups identified in the case studies fall into this category. For example, the virtually self-supporting group employed for composer assembly in Case M2 operated largely as a non-mechanical flow line, as did the seven-man television-assembly group in Case L. These examples suggest that whilst in certain conditions there exists the need to divide the total work content of the job between workers the resulting flow system does not prevent the adoption of an effective work organization providing tasks are not highly rationalized, or product-flow highly intensive.

Clearly, moving-belt-type lines, whether of the fixed-item or removable-item type, in general provide less favourable conditions for the creation of formal functional work groups. The use of large cycle times on fixed-item lines, and the reduction of the mechanical pacing effect by variation of line speed, item spacing and station parameters, have precisely the same effect as increases in cycle time and the increase in buffer capacities on non-mechanical lines; however, practical considerations, such as product size, may well limit the amount of freedom available to the individual. Such an approach might be combined with line division in order to create some degree of group working.

For example, the division of the indexing line used for engine assembly in Case D, perhaps only by the provision of an empty station at certain points of the line, might have provided conditions in which some self-organized group working was possible. Whilst buffer stocks can be provided on removable-item, moving-belt lines,[4] their use is unlikely to substantially reduce pacing since a worker must be present at a station to take items from the line into stock, hence the reduction of pacing is again largely achieved by manipulation of line speed, item spacing and station parameters.

Line division and station grouping are perhaps more appropriate for moving-belt lines; indeed, many of the examples of functional group working on moving-belt lines rely upon such an approach.

This chapter has concentrated upon the creation of conditions appropriate for the existence of formal functional groups. Little reference has been made to consultative groups; however, it is clear that in some respects the prerequisite conditions for functional groups apply also to consultative groups. In particular, group size, spatial proximity and freedom for interaction are likely to be important for effective consultative group operation, since without these it is unlikely that there will be sufficient cohesion to permit group consensus on issues, such as the nomination of representatives, or opinions on topics to be discussed with management. Of course some of these prerequisites may be provided, to some degree, by facilitating 'off-the-job' interaction between workers. Social areas and meeting rooms may be provided, as in Case A1.

Additionally, or alternatively, formal programmes may be established in which groups of workers meet, perhaps during working hours, for the purposes of discussing problems, receiving extra training, etc. The creation of such consultative groups may provide some degree of group identity and interaction in conditions which prohibit functional group working. Equally, the existence of functional groups may obviate the need for separate consultative groups, since the assumption of autonomy and responsibility will often be accompanied by the need for joint decision making between the group and management.

References

1. Sirota, D., and Wolfson, A. D., 'Job enrichment: surmounting the obstacles', *Personnel* (July/August 1972).

2. *Work Structuring: A Summary of Experiments 1963–8*, Philips Co., 1969.

3. Buxey, G. M., 'Assembly line balancing with multiple stations', *Management Science*, **20**, 6 (February 1974), pp. 1010–1021.

4. Franks, I. T., Gillies, G. J., and Sury, R. J., 'Buffer stocks in conveyor-based work', *Work Study and Management Services* (February 1969), pp. 78–82.

CHAPTER 10

Formal Work Groups in Other Mass-production Systems

In examining the extent to which group working may exist within the context of individual assembly, collective working and automated assembly, some of the operating characteristics of these systems will be studied, and examples provided.

INDIVIDUAL ASSEMBLY

In theory the concepts of group working and individual assembly might appear to be mutually exclusive, but in practice this would not seem to be the case, as will be shown below.

Basic design considerations

Unlike flow-line working, individual assembly does not require the arrangement of workers in close proximity. Although in practice workers will often be grouped together when engaged on like or similar items, this may not apply when dissimilar items are being assembled. Equally, unlike flow-line working greater freedom will exist in the arrangement of operationally independent workers, hence generally it should be possible to provide the physical grouping which is usually considered to be a prerequisite for group working. Furthermore, the independence of workers should provide greater freedom in the determination of group size.

Worker interdependence is unlikely to be a requirement of such groups, nor are workers likely to see their skills and tasks as complementary, although some degree of interdependence may be introduced in all such groups, as will be shown below. Individuals engaged on the complete assembly of an item may find it easier to identify themselves with the product and feel a sense of completion. Such individuals are also able to regulate their own activities, and may be judged by their own results, since clearly defined task boundaries exist and performance criteria are available. However, both closure and self-regulation apply in this case to the individual directly rather than through his membership of a work group. For these reasons integration within the group may be minimal although certain of the conditions for group cohesion exist.

118

Both cohesiveness and integration may be higher when individuals are engaged on the assembly of different items since there is then some differentiation between members and some evidence of complementary activities. Furthermore, such arrangements provide greater opportunity for the allocation of group autonomy and responsibility.

Autonomy and responsibility

Groups of individual workers assembling similar or different products might assume some degree of autonomy in respect of both quantitative and qualitative goals. They may as a group decide the allocation of models or items to individuals, and may, if working against given output requirements, determine batch sizes and short-term schedules for each model or item. The distribution of tasks within the group will in general be determined by the allocation of models or items to individuals; however, even within a fixed production schedule workers may choose to rotate between models or items whether or not rotation and work allocation is determined by the group. The operational independence of workers should permit some individual flexibility of work hours, breaks, etc.

In an extreme case groups of workers engaged on individual tasks may choose to adopt some form of work colaboration, perhaps even resulting in a form of flow-line working. Some cooperation between workers may be desirable to ensure full utilization of equipment and to avoid worker idle time. Equally, certain tedious jobs may be rotated, whilst indirect and intermittent responsibilities such as packing, cleaning the work area, paperwork and liaison with other departments might be undertaken by members in turn, or by a representative of the group. Equal division of such responsibilities will be necessary when individual-incentive systems are employed, whilst group-incentive or day-work systems permit greater flexibility of operation.

Training may be facilitated if the group are allowed to allocate models and tasks, since then some division of operations may permit the trainee to learn parts of the job before taking on a complete assembly operation. Furthermore, the different parts of the total operation may be learnt from the different individuals in the group who are recognized to be most proficient.

The degree of autonomy of groups of workers engaged on the assembly of identical items is limited since there is no possibility of model item or task alloction other than the division of operatives to provide a form of flow-like working. Some self-organized rotation may be appropriate, but in this case the purposes are likely to be largely social.

Operating characteristics of 'individual' work groups

Individual working can be considered an extreme case of non-mechanical flow-line work, hence the operational characteristics identified in the previous chapter may be cited here. In comparison to flow-like work, individual assembly benefits from zero balancing loss, and, providing adequate material and parts supply is provided, no system inefficiency should be incurred. Equally, since the total work content of the assembly is not subdivided, some non-productive work is avoided. Offsetting these advantages are the probable need for more training, increased space, equipment, and work in progress.

Individual assembly within the context of an autonomous functional work group provides not only the benefits cited above, but may also facilitate work balance. Often, particularly when output requirements are fixed as in the case of subassembly work, the time needed to assemble the requisite number of items in a period, will not correspond exactly to the time available from available labour, hence some inbalance may occur. The possibility of cooperation between workers engaged essentially on individual work provides one means of overcoming such inbalance.

Examples

The examples below, two of which are taken from the studies in Appendix E, illustrate radically different approaches to the use of individual working with a work-group organization.

Example 1: Philips Company[1]

In the original situation, three foremen, fifteen chargehands and two-hundred and thirty-one assembly workers and inspectors worked under one departmental head. The work, largely process-orientated, light assembly work, was performed mainly by females. Cycle times ranged from five to one-hundred and twenty seconds and each assembler had her own job, the foreman being responsible for allocating the work, etc. Payment was by individual incentive.

The following changes were introduced:

(1) The formation of independent groups each responsible for job allocation, materials supply, quality inspection and submission to final inspection, and appointment of representatives.
(2) The job of the foreman was eliminated and chargehands were made group leaders with wider responsibilities for consultation and worker assessment.
(3) Sample inspection was reduced.

Example 2: Assembly of television control unit (Case N—Appendix B)

The remote-control device for domestic television is assembled individually by each of approximately twelve female workers. Total work content is approximately twelve minutes, six distinct operators being required in the assembly of each of the identical products.

Each worker begins preparation of a printed-circuit board at one work bench, and moves to the next bench to finish the preparation of the board. Insertion of components by the same worker is undertaken at a further bench, then, following soldering by a male worker, work continues at three further benches with the assembly of push buttons and springs and finally the assembly of the case for the unit.

The individual assemblers in the group are together responsible for developing a procedure which avoids delays due to waiting for access to a particular work bench which in turn influences their choice of batch size, etc. They are together responsible for determining their own output, payment being on a day-rate system.

Example 3: sub-assembly work for composer unit (Case M2—Appendix B)

Approximately fifteen male and female workers, part of a larger group of thirty assembly workers, are engaged on the assembly of different subassemblies for a product which resembles an electric typewriter. These workers are together responsible for undertaking approximately eighty operations ranging in cycle time from one minute to one hour. They work to a weekly output schedule and within this target are responsible for allocating work between themselves. They are required to deliver all completed subassemblies to stores, complete their own paper work, etc., and were also instrumental in redesigning the layout of their work area.

COLLECTIVE WORKING

The term *collective working* is used to describe a production system in which workers engaged on assembly work, work together on one product. Whilst such an arrangement may exist on a flow line, in the manner identified in the previous chapter, we are concerned here with collective working as an alternative to flow-line systems. However, the prerequisites that are identified apply also to collective grouping on flow lines.

Collective working provides an alternative system providing items are of sufficient size to permit workers to be employed together in one area and providing excessive worker idle time through work interference can be avoided. The avoidance of interference depends upon

(1) The nature of the precedence constraints governing the work elements of the total assembly task. The absence of any sequential dependence amongst work elements will ensure that workers do not incur idle time through having to wait for one of their colleagues to complete a task before being able to continue their own tasks.
(2) Since in practice complete independence of elements is unlikely, the avoidance of work interference may also depend upon the allocation of tasks to workers. If tasks are allocated to workers such that their operations are independent then, even though some of the constituent elements are sequentially dependent, work interference is avoided.

It should be noted that the tasks must also ideally provide for the equal division of the total work content between workers. However, because of the indivisibility of elements, such perfect balance of operations is not usually possible. Furthermore, the greater the constraints on the allocation of elements to workers imposed by the need to avoid interference, the less the likelihood of a good work balance.

Imperfect work balance can, however, be overcome to some degree through worker cooperation, i.e. workers with shorter operation times may, on completion of their operations, assist those workers whose operations are not complete. In practice, because of physical limitations at the work-place, limitations on tools, and non-productive work associated with task changes, workers cooperation will be less than perfect, hence some degree of imbalance will probably occur.

Subject to the above limitations, a form of collective working may be employed in which more than one item is available to workers. Thus a worker having completed his operation on one item may move immediately to a second item. In such a case two (or more) items will always be available to workers in the system. An item will leave the work area when all workers have completed their operations on it, to be replaced as needed by a further item for assembly. This availability of more than one item provides the opportunity to reduce potential imbalance, by

(a) Providing greater opportunity for worker cooperation, i.e. workers with shorter operations may find it easier to assist their colleagues having accumulated 'spare' time over several items.
(b) Work may be allocated to operatives in such a way that they, for example, perform certain elements on some, but not all, items thus facilitating overall work balance.

FORMAL WORK GROUPS IN COLLECTIVE WORKING

Basic design considerations

Work-group cohesion will be facilitated by the proximity and interdependence of workers, and by their common goal, whilst the logical grouping of workers and the possibility of them together completing a significant module of work may provide the 'closure' which is generally advocated. The size of the work group is likely to be severely limited because of the physical limitations of the work-place. The group should not therefore be too large to prevent consensus or effective communication, although in a small group interpersonal relations are likely to be important and a change of membership may have a substantial effect on the nature of the group.

The similarity of the tasks performed, and the interdependence of workers, should ensure that members' tasks and skills are seen to be complementary and, furthermore, all members are likely to be able to appreciate the skills of their colleagues. Communication with peers is facilitated by the physical proximity of the members of the group, whilst a form of group payment is practical providing approximately equal work is performed by members. The clear task boundaries of the group, and the possibility of establishing precise performance criteria based on output, provide the basic conditions necessary for the group to be able to regulate their own activity.

Autonomy and responsibility

Autonomy in respect of the distribution of tasks may be provided as the group may allocate operations to members, and provide cooperation, perhaps even alternating certain tasks. In some cases, particularly when tasks are independent and when a work balance is readily achieved, this autonomy may involve little extra work for the group. However, the avoidance of work interference, excessive imbalance and high worker idle time may depend far more upon the internal organization of the group, especially if it is recognized that the performance of members of the group may vary, that external disturbance may affect task times and that new workers may require quite a long training time to reach an acceptable performance level, all of which may necessitate short-term reallocation of tasks, etc. Furthermore, the assembly of more than one model or item, the need to adjust equipment, replenish supplies, clean the work-place, liaise with other workers, complete paperwork, and all such 'irregular' tasks will probably necessitate some short-term reallocation of tasks or cooperation, if idle time is to be avoided and output is to be maintained.

Such a group may also undertake work-method improvements, select internal and external leaders and representatives, and train new members. If collective working groups are employed on the assembly of complete items or subassemblies, it is likely that they will either be, or can be, insulated from subsequent stages of the production process by buffer stocks, and hence autonomy in respect of work-hours is more easily provided. The group might arrange their own overtime working, fix their own breaks, lunch hour, etc. and may even determine their own daily work-hours within certain constraints, as for example in a 'flextime' system.

Operating characteristics of 'collective' working

A comparison of the operating characteristics of collective working systems and non-mechanical flow lines is presented elsewhere.* This comparison through digital-computer simulation demonstrates the advantage of collective working in respect of system output (or conversely idle time), work in progress and the space required by the system to accommodate work in progress, i.e. the system capacity. For a given total space requirement, and given number of workers, the collective-working system operates with less idle time. Alternatively, for a given output a flow line requires far more space and will accrue considerably more work in progress than an equivalent collective system. Hence, although many of the advantages of collective working are of a behavioural nature, the system offers the benefits of either lower work in progress or higher output efficiency in comparison to a conventional flow line. As these results were obtained from the simulation of collective working without worker cooperation, these relative advantages would be greater in practice since cooperation would not only facilitate better notional balance in the system, but also help minimize idle time due to system loss (i.e. short-term imbalance between workers).

*Slack, N. D. C., and Wild, R., 'Production flow lines and 'collective' working—a comparison,' *Int. J. of Prod. Res.*, forthcoming.

AUTOMATED/MECHANIZED SYSTEMS

The case studies in Appendix B give examples of assembly automation corresponding to lower levels identified in Chapter 4. Motor-vehicle-engine assembly in Case D provides an example of work 'in line' with a synchronous-indexing, automated-assembly system, although in this case, because of a reduction in the extent of mechanization on the line, the mechanized stations could be more truly considered to be 'in line' with the manual system. Several examples of ancilliary work are given, typically involving the loading and unloading of a single-station or rotary-indexing assembly machine whilst the linking together of such machines provides for similar work with interdependence of operations. Examples of a slightly different type of system can be found in printed-circuit-board assembly.

It is within the context of these types of semi-automated, or mechanized, systems that we shall examine group working. Doubtless it would be easier to provide for formal group working in more advanced systems; indeed, examples exist of such a method of work organization being used in automated continuous-process systems. However, it is evident from the discussion in Chapter 4, and the observations reported in Chapter 6, that automation in assembly work, indeed in most complex discrete-item mass production, differs substantially from process automation. In fact, we have seen that in most cases increased automation of such production leads initially to a deterioration in the direct work roles of the operatives. In view of this effect, of what in many cases is a necessary development, we should perhaps look to alternative forms of work organization to provide some compensatory effect for workers. Even the most constraining type of system can perhaps be deployed in such a way as to facilitate the use of formal work groups. However, before looking at such possibilities, we will look briefly at the basic system as described above.

SIMPLE MECHANIZED WORK

Basic design consideration

Workers 'in line' with multi-station mechanized systems are likely to experience much the same constraints as workers on short-cycle-time, moving-belt flow lines. Work on indexing machines can perhaps be improved, that is, the pacing effect may be reduced, if operator controlled or non-synchronous indexing is employed; however, even in such cases, the worker is likely to have little opportunity to leave his work-station unless replaced by another worker and, of course, his regular work-contact with other operatives will generally be restricted to adjacent stations unless these also happen to be mechanized. A similar, and in some respects worse, situation applies to the operators engaged on load and unload tasks on a single-station or rotary-indexing machine. Here the social isolation may be worse since it is unlikely that more than one such worker will be associated with each machine. Freedom to leave the work station will exist only if the machine is operator controlled, although in such cases the freedom may be greater since other workers may not be employed on the machine. In this respect work on printed-circuit-board component assembly tables, provides an example in being manned and controlled by a single operator, thus allowing some freedom but affording little social contact during work periods.

Autonomy and responsibility

In machine-controlled systems, functional group working is virtually impossible. Some degree of job rotation may be possible even though the 'exchange' of jobs may have to take place when machines are stopped, e.g. before start-up or at break or lunch times. This in turn may mean that rotation must be organized. The fixed nature of the operations to be performed rules out autonomy in respect of task distribution, whilst the machine pacing of the work ensures that the extent to which workers can control work speed, and therefore determine schedules, is limited. Such a possibility may exist, however, together with the opportunity to arrange task distribution and rotation, if the number of workers is less than the number of tasks to be performed. However, in the context of machine operation and in-line work, this situation is unlikely since in general relatively expensive equipment will be fully utilized. Operator-controlled systems provide for greater worker freedom and thus it becomes feasible to think in terms of self-organized job rotation and performance of direct and intermittent duties such as supply of materials, sampling inspection and completion of paperwork. Such autonomy is more readily discharged when workers in a group are operationally independent as, for example, in a group of workers each associated with single-station or small rotary-indexing machines. Interdependence of operations, even though operator controlled, will probably restrict autonomy to job rotation where very large buffer stocks of items can exist between stations.

Worker responsibility for machine setting, maintenance and repair may contribute to group autonomy in both operator- and machine-controlled and operationally dependent and independent groups. Such tasks may provide the opportunity for cooperation, especially in operationally dependent systems, whilst in independent operator-controlled systems cooperation may be employed, or alternatively one worker may be assigned as a 'specialist' for the indirect work on one machine, or such tasks may be undertaken on a rotation basis.

Examples

The first two examples below relate to metal working but nevertheless demonstrate how functional group working might be employed in the context of mechanized functional systems.

Example 1: forging of automobile parts (Case E1—Appendix B)

Three workers are engaged respectively on an induction furnace, a forge and a press, connected in this order by roller conveyors. The group together manufacture steel blanks for a suspension link-arm for cars. Each piece of equipment whilst mechanized is operator actuated. The loading and unloading of each machine is also manually controlled. There are no buffer stocks between machines. This group of workers rotate jobs on a self-organized basis; however, each man is considered to be a specialist on one machine, hence special work on each machine, e.g. setting or repair, is undertaken by one man only.

Example 2: machining of bracket (Case E—not reported in Appendix B)

The machining of brackets for truck brakes provides an extreme case of functional group working in the context of a largely mechanized system. Three workers together operate seven machine tools. The large buffer stocks between machines, the comparatively long cycle time at machines (approximately three minutes), together with the fact that machines are operator controlled or activated, provides the operator freedom necessary for the assumption of considerable group autonomy. Following the proposed reorganization of this group, the three workers will be given responsibilities for job rotation, the allocation of operations to members, quality inspection and rectification. The group is to be responsible for the requisitioning and charging of tools, and will be paid on a group incentive system.

Example 3: valve-finishing and testing (Case 23 and 24—Exhibit 6.1)

Workers engaged on ancilliary (loading and unloading) and machine-minding tasks on sections of a largely mechanized valve-assembly system were given some of the responsibilities previously discharged by their supervisors. The workers, who were able to leave the system for short periods of time, were operationally independent, hence they were able as a group to undertake certain indirect and intermittent tasks, including responsibility for quality inspection, supply of materials, completion of paperwork and adjustment of equipment.

The testing of the valves produced by such work groups is undertaken by pairs of workers, each associated with a semi-automatic test machine. The pairs of workers, who rotate jobs, are together responsible for loading/unloading the machine (a machine-paced job), all service tasks, manual retesting of valves, completion of paperwork and maintenance of all records.

INTEGRATION OF OPERATIONS

The creation of product rather than process-oriented departments as production sections provides increased opportunity for effective functional-group membership for workers engaged on mechanized systems. Often, even in mass-production plants, a form of process layout is employed for mechanized or semi-mechanized processes, especially when such processes are used for the production of subassemblies and components. For example in electrical engineering, processes such as coil winding will normally be found in a separate department. In such cases a large number of machines and operators performing similar work will be located together. A similar situation often applies when single-station or small rotary-indexing machines are used for the manufacture of subassemblies. In such cases, even though machines may be operator controlled and operationally independent, the scope for functional-group working may be limited since all workers may be engaged on similar jobs. Thus the benefits of job rotation and the scope for group allocation of tasks are minimized. Furthermore, such work—which has many of the characteristics of individual assembly of identical or similar items—is unlikely to satisfy the basic design considerations for cohesive integrated group working.

An alternative arrangement, perhaps more appropriate for small single-operator machines, provides for the location of the mechanized process alongside the other processes required for the manufacture of a complete item. Such a *product arrangement* may be particularly beneficial for the worker on the mechanized equipment. A logical grouping of workers is formed, where a sense of closure is facilitated. Group interdependence is emphasized, complementary skills are deployed, whilst the use of a group payment system may be appropriate. The wider variety of jobs provides greater scope for job rotation and task allocation, although operator freedom from his machine is still required for effective group working.

Examples

These two examples demonstrate how mechanized jobs which would normally be undertaken in a separated area may be integrated with other operations required for product assembly. In the first case this integration is not provided for the purpose of group working. The second example describes a situation in which group working is of paramount importance.

Example 1: assembly of bells and chimes (Case 9—Exhibit 6.1)

Assembly workers in this plant are arranged in completely self-contained units. Each unit, consisting of approximately twelve to fifteen workers, undertakes the complete assembly of a comparatively simple product. No subassembly areas exist hence the electrical coils, required in most products, are also wound, terminated, etc. in the production units. Each such unit, is fed from a coil-winding machine manned by one operator.

Example 2: television assembly (Case L—Appendix B)

A group of seven workers were together responsible for the complete assembly of black-and-white domestic television sets. The operations performed by the group included the production of printed-circuit boards, i.e. board preparation, component insertion, wire crop, flux and solder. Two workers were engaged on printed-circuit-board work using indexing assembly tables and automatic soldering equipment. The group operated at a cycle time of approximately twenty minutes; they were responsible, however, for setting their own output targets, for their own quality control and rectification, for allocation of tasks, job rotation, requesting materials and liaison with other departments. Thus the circuit-board workers whose equipment was all operator-controlled were able to rotate within the group, as well as undertaking indirect tasks, etc.

GROUP TECHNOLOGY AND THE 'CELL' SYSTEM

Whilst not fundamentally concerned with mass production, group technology can be considered as a means to provide for the mass production of the similar components needed for batch-produced products. Conceptually, therefore, group-technology production has similarities with mixed and multi-model flow production. Some similarity is perhaps also evident in the relationship of the *cell system* of organization in group technology and functional group working in flow-line production—indeed, recent discussion of the cell system has pointed to the behavioural benefits of this method of organization whilst citing the adoption of group working in flow-line production as an application of the same principle. This argument, identified in Chapter 3 as a further possible indication of the benefits of group working in production. is examined briefly below.

Group technology attempts to apply some of the benefits of mass production to the batch production of products. This objective is pursued through the grouping of parts or components to provide families of items which are sufficiently similar to permit the use of a flow, but not necessarily linear, type of processing system. The term 'cell-system' refers to the organization of men and machines in this situation. A cell is therefore a group of facilities associated with the manufacture of usually one, or perhaps more, group or family of items or components.

The creation of manufacturing cells appears to differ in several fundamental respects from the creation of functional work groups in mass-production systems. The objectives of group technology will ensure that work in progress and item-throughput time are generally reduced. The variety of work undertaken at any machine may be reduced, whilst batch size will probably be increased and thus the need for machine setting reduced. Such changes in effect impose some of the constraints of conventional flow-line working on workers who had previously operated individually. However, as it is often impossible to balance facilities in cells some labour flexibility is required. Flexibility facilitates job rotation, one benefit of which is to provide the worker with the opportunity to deploy more skills in the operation of a variety of machines rather than being confined to a single type of work at one machine. Thus whilst the creation of cell systems leads to the provision of less intrinsic work variety at each machine, together with certain of the detrimental characteristics of flow lines, e.g. closer work-pacing, the normal need for some labour flexibility provides some compensatory social interaction and interoperation variety.

The product grouping of facilities may provide greater closure and worker interdependence whilst group payment may be appropriate, skills may be seen to be complementary and objectives shared. Clearly, many of the conditions for the existence of a cohesive integrated work group exist, although the structure of the group and the degree to which the group might assume autonomy may be limited by the size and degree of mechanization of the equipment.

In this respect the nature and extent of functional group working possible in cells is limited in the same manner as in certain types of mechanized assembly. Indeed a group-technology cell can be considered as an integration of operator-controlled mechanized operations of the type advocated in the previous section. The creation of such cells therefore provides the opportunity to create, but does not implicitly result in a form of semi-autonomous group working. As with other mechanized systems, this opportunity is dependent upon worker control of, and freedom from, machines, and some degree of operational independence between workers.

In such conditions workers in a cell can together assume some responsibility for the allocation of tasks and for job rotation. The group may schedule work and determine batch sizes against given output requirement. Some indirect responsibility, such as materials supply and the completion of paperwork, may be undertaken; indeed the group may take on certain of the responsibilities of supervision and technical functions providing freedom exists within the group. Since one objective in the establishment of such cells will generally be the reduction of work in progress and throughput time, it is unlikely that sufficient independence will exist for operators to determine their own work-hours or breaks; indeed such independence might more easily be provided in the type of independent working which in general is replaced by the cell system. However, the creation of relatively self-contained groups may permit some group flexibility of hours.

Clearly the cell system in batch production provides the opportunity for the creation of semi-autonomous functional group working,[2] providing labour flexibility is available and machines are operator controlled. Indeed it might be argued that, as with mechanization of assembly work, such work organization is necessary if workers (in this case perhaps skilled workers) are to be compensated for their otherwise diminished work role.

DISCUSSION

It is clear from the discussion in this and the previous chapter that each of the systems identified permit to some degree the adoption of formal functional group working. In no case was it found to be impossible to visualize the existence of at least some of the characteristics of group working, although equally it was clear that in most cases the nature of such groups would be influenced to a very considerable degree by the system in which they were to exist. Even in the most limiting situation it should be possible to provide for the satisfaction of many of the basic design requirements of effective work groups; however, group structure, cohesion and integration do not in themselves promote autonomous responsible group working. Thus, whilst even rationalized and paced flow lines, and machine-paced automatic assembly systems, might support cohesive groups of workers, the degree of autonomy and responsibility influences the extent to which they are able to operate as self-organizing and effective production units capable not only of fulfilling organizations goals, but able also to provide the structural requirements for satifying jobs.

In these two chapters we have examined the effect of the constraints which derive from the nature of the production system. Despite the significant differences between the type of production system examined, the constraints identified bear a close resemblance. In fact, in virtually all cases they can be expressed in terms of their effects on the 'freedom' of the individual, i.e. his ability to leave his immediate work area or machine during working periods, which in turn is influenced by the degree of work-pacing and machine control of operations and the interdependence of workers.

Thus highly rationalized machine- or operator-paced lines and machine-controlled operator-interdependent mechanized systems afford little worker freedom. In these systems, whilst formal functional work groups may be established (e.g. on divided or station-grouped lines), little scope exists for the assumption of autonomy in respect of work goals, task allocation or performance. Thus effective job restructuring is precluded and the need for external organization is maintained.

It appears that without the individual freedom which is associated with low pacing, operator interdependence and worker control of equipment, the autonomy of work groups is limited mainly to those dimensions which are not directly connected with the nature, planning and control of the work done by the group, i.e. non-functional dimensions. Whilst some degree of self-organization in these areas may well be beneficial, it seems unlikely that the principal benefits of group working identified in Chapter 8 would be afforded without functional autonomy.

Whilst not specifically examined in these chapters it is evident that constraints limiting the nature of group working derive not only from the production system but exist within the system and apply to it. Constraints are applied to the system as a result of its relationship with both preceding and succeeding stages of the production process. Thus if a succeeding stage is closely dependent on the system there will be little opportunity for the selection of production schedules. In fact without some degree of independence of systems, perhaps provided by buffer stock or duplication of systems, many of the benefits of self-organization and system flexibility are lost. (Such a situation existed in company C where, because of external constraint, the flexibility of individual assembly could not be utilized during periods of absenteeism.)

Constraints within the system may prevent the exercise of group autonomy and responsibility even though the system provides it. For example, unless workers are sufficiently flexible, i.e. able to perform several jobs, task allocation and job rotation will be limited. Similarly, internal constraints may stem from payment and skill differentials within the group, relations with supervision and other departments, and the layout of facilities.

References

1. *Work Structuring: A Summary of Experiments at N.V. Philips*, Philips Company, 1969.
2. N.E.D.O., *Production Planning and Control*, H.M.S.O., 1966.

Part 6
The Development of Mass-production Systems

A reexamination of current and likely future developments of discrete-item mass production systems, and the formulation of principles for systems design and development.

CHAPTER 11

The Present Situation and Future Developments

In this chapter we shall attempt to evaluate the present and speculate on the possible future practice in the mass production of complex discrete items and in doing so we shall reexamine some of the points sketched out in Chapter 3 in the light of the information developed in the intervening chapters. The objective throughout will be the identification of factors influencing the selection of systems and their effectiveness, with particular reference to the pressures which may give rise to the need for changes in present practice.

CURRENT DEVELOPMENTS

It was hypothesized in Chapter 3 that manual flow lines currently provided the principal system for the mass production of complex discrete items. This hypothesis was supported by the observations made during the survey summarized in Chapter 6. It was further hypothesized that the principal dimensions in the current development of such systems embraced the use of higher levels of automation/mechanization and the provision of less-constrained manual work. Again the observations in Chapter 6 appear to support this hypothesis. To some extent, however, the reasons for and effects of such developments were found to be contrary to prior expectations. Whilst the behavioural disadvantages of manual-flow-line production (discussed in Chapter 3) have been the subject of considerable and increasing interest, many cases that might appear to be job-restructuring changes aimed at overcoming some of these disadvantages have in fact been changes prompted by production and engineering considerations. Furthermore, whilst increased automation has been widely advocated as a means to eliminate the boring alienating jobs which flow-line work is commonly considered to have created, such development in this field of mass production appears in general to have led to the deterioration of many of the remaining manual jobs.

Recent developments in the area might be summarized as follows:

(1) Restructuring and organizational changes have in general given rise to the rearrangement of assembly lines, through, for example, reduction in line length, increase in cycle time and the replacement of lines by individual

working. Individual assembly has relied upon the enlargement of tasks, and the removal of pacing effects, whilst group working has generally been introduced in flow-line production, the benefits deriving mainly from system flexibility through the allocation of increased responsibility and autonomy to the group.

(2) The motives for changes have often related to the need to increase system flexibility to accommodate absenteeism, product and demand changes, disruption of supply, etc., and the need to increase or ensure product quality. Changes have normally affected systems devoted to the assembly of comparatively small items. Restructuring and organizational changes relating to work normally performed on moving-belt, fixed-item lines have proved more difficult and such situations, e.g. the motor industry, are now increasingly the focus of attention. A switch to individual assembly methods has by and large been the result of a need to compensate for changes in either the product cycles or the nature of the production-line worker and/or the general labour pool. Group working has also been adopted as a means to combat these changes, and in addition the possibility of worker interaction and cooperation has proven valuable in overcoming problems such as line balancing and model sequencing, and as a method of providing job enrichment.

(3) Faced with steadily rising labour costs, some industries have increasingly turned to automation and mechanization. Additional potential advantages have been seen to include improved product quality, the removal of operators from hazardous operations and a reduction in floor space required for assembly. Mechanized assembly has been introduced in industries where conditions are particularly favourable, i.e. large-quantity production and high production rates with low product obsolescence risk over long periods of time. These conditions facilitated the application of relatively expensive, inherently inflexible special-purpose machinery. Where product life-span is shorter, but predictable, initial capital outlay is more important, hence mechanized assembly has, of necessity, taken on a more general-purpose nature. The trend has been to make use of modular construction with configurations of largely standard units. A recent development has been the use of industrial robots, 'easily programmable, operator-less handling devices that can perform simple, repetitive jobs that require few alternative actions and minimum communications with the work environment'.

(4) Mechanized assembly commonly integrates manual and automatic operations on a mechanized indexing-transfer line and workers have often been retained for operations which cannot be economically mechanized.

Public opinion

Media coverage of many of the topics dealt with in this book has never been so high as in recent years.[1,2] Thus public opinion has been heightened and general discussion of such topics is widespread. Whilst the dissemination of information can only accelerate changes—if changes are seen to be required—there is a danger that wide publicity may help establish certain misconceptions which may in effect promote resistance to change. Such misapprehensions appear to be prevalent in respect of group working and the nature of flow-line work. Group working is now often advocated as one solution to the problem of flow-line work, and is generally seen as an alternative production system. Thus the real possibility of providing enriched and motivating work, together with the use of flow-line production systems, is often overlooked. Widespread reference to the work undertaken at Volvo and Saab appears to have given the impression that these companies have abandoned flow-line work, whereas in fact their distinctive achievement is not their abolition of the assembly line, nor simply the enrichment of assembly work, but the provision for job enrichment within the context of flow-line work. This focus on the supposed need to abolish flow-line work, which stems perhaps from fundamental misinterpretations of certain well-publicized exercises together with a misunderstanding of the nature of group working and a failure to distinguish between individual job and organizational changes, may have the effect of exaggerating the difficulty of the solution of the problems which have been discussed. This in turn may discourage companies from either recognizing such problems, should they exist, or in looking to their solution.

International precedents

Much of the recent discussion of flow-line and mass-production work has been prompted by reports of exercises undertaken in other European countries. It was pointed out in Chapter 3 that whilst the results of studies in such countries might be of relevance in the U.K., countries such as Sweden and Norway differ substantially from others (especially the U.K.), hence it might be innappropriate simply to attempt to copy the type of exercise undertaken in companies such as Saab and Volvo.

Some of the societal, cultural and organizational factors which distinguish the U.K. from Scandinavian countries have been examined in detail elsewhere,* where the emphasis was on the identification of those factors which might influence the appropriateness and effectiveness of job restructuring and work-organization changes. Both the creation or emergence of production problems, and the manner of their solution, were found to be possibly influenced by similar cultural factors. It was concluded that the following seven specific cultural factors might be considered to be important in the restructuring of jobs and work reorganization—and that these factors might be seen as prerequisites for effective change, and as dimensions distinguishing the U.K. from the Scandinavian countries.

* Birchall, D. W., Elliott, R. A., and Wild, R., *Organizational, Cultural and Societal Differences and their Effect on the Application and Effectiveness of Job and Work and Organizational Changes*, Administrative Staff College, Henley, 1973.

(i) A positive and cooperative attitude on the part of the workers to change.

(ii) The workers must have a strong desire for participation and accept its legitimacy.

(iii) An instrumental orientation to work should be absent, or should be amenable to reorientation.

(iv) The management should be 'humane', forward thinking and have a belief in the legitimacy of worker participation.

(v) A healthy industrial-relations mechanism must exist for negotiation and consultation.

(vi) Industry should have an organismic or flexible organizational structure.

(vii) There should be high educational and living standards.

It can be argued therefore that the conditions prompting both the need for action and the nature of the action taken in countries such as Sweden and Norway may differ from those presently existing in other countries, e.g. the U.K. This hypothesis was to some extent substantiated by observations during the studies discussed in Chapter 6. However, the argument that 'what is now happening in Scandinavia, will shortly happen elsewhere' is not necessarily invalidated by such observations, since conditions in other countries may change. The factors listed above do nevertheless suggest that change in another country may not *in itself* be sufficient justification for similar changes elsewhere.

PRESSURES AFFECTING CHANGE

Various factors which might conceivably influence the choice of production systems both now and in the future, were outlined briefly in Chapter 3. The discussion below takes up some of what now appear to be the more pervading and important issues. This brief discussion should provide at least some basis from which to speculate on the future selection and use of systems. Some of the possible consequences of these pressures will be identified prior to the presentation of a speculative view of future practice and development in the mass production of complex discrete items.

Work needs and attitudes

Throughout this book, because of our essentially pragmatic approach, we have been able to avoid the examination of certain topics which it might be argued are central to the issues being evaluated in this chapter. It has been argued that the highly rationalized and constrained manual work which is characteristic of many assembly-line systems is incapable of providing the self-fulfilment considered to be important by the present generation of workers. If this argument could be substantiated, it would provide a very considerable pressure for the need to change present production methods. However, despite substantial research, it is not generally proven that such work itself necessarily gives rise to dissatisfaction or alienation, that self-actualization is a dominating motive amongst workers, or that job dissatisfaction is necessarily dysfunctional for the organization. We must, however, dismiss the last item mentioned, since it is surely no longer acceptable to argue that productivity alone is a reason for a company to seek change.

The aspects of jobs and the nature of the work method and content which are most likely to influence feelings of self-fulfilment and satisfaction have been the subject of considerable research, discussion and controversy. We have attempted to avoid entering into such a debate by seeking a consensus view on the subject. (It happens that the desired job and work attributes developed in Chapter 5 cover many of the areas of 'motivation' proposed by Herzberg[5] whilst excluding virtually all of his 'hygiene' factors.) This rather pragmatic approach was considered to be preferable to one invoking one or more of the present theories of job enrichment, etc. However, the establishment of a rationale for change provides no proof of the needs for such action. Fundamentally, such proof rests with the examination of the nature of the job needs and attitudes of workers—another area which has for some time been characterized by sharply differing views, largely because of the difficulties in understanding the complex interaction of the very large number of variables involved.

Despite this conflict of views, and the contradictions existing in many recent research findings, one important issue is evident and will be described here in terms of certain recent research studies.

Goldthorpe,[6] studying workers in the motor industry in Luton, looked at their decisions, particularly their choices of job, in order to obtain an insight of their needs. He found that workers in the motor industry had *sought* such jobs, often having given up more interesting jobs elsewhere in order to move onto the vehicle-production lines. They attached greatest importance to pay, and this was why they chose to remain in such jobs even though others were available to them. They described themselves as being satisfied with their jobs despite the nature of the work they were engaged upon, because of the balance which existed between their dominant needs and what their job provided. They apparently did not seek approval, acceptance or recognition, nor did they seek membership of work groups, since theirs was an *instrumental* orientation to work. Daniel,[7] examining these findings, challenges the concept of an 'orientation to work' and suggests that a different set of priorities may relate to different situations. He points out that it is not so much what a worker is interested in that is important, but rather which aspects of all the things that he is interested in are most important in particular situations. Citing various cases he shows that whilst workers may press for changes which reflect an instrumental orientation to work, e.g. improved pay, hours, conditions, etc., this does not preclude the possibility of them subsequently enjoying changes associated with job enrichment, such as increased discretion, variety, etc. In a negotiating situation this may mean that workers resist aspects of job restructuring whilst pursuing improvements in work load, pay, etc. However, despite this they may, following implementation of agreed changes which involve some restructuring of his job, express proportionately greater satisfaction with these aspects. Wedderburn[8] points out that a worker's experience at work may contribute towards orientations to work, and shows how for most manual workers the dominant experience is their relationship to the work itself.

This may explain two apparently paradoxical situations. Firstly, the fact that in attitude surveys amongst workers engaged on flow-line-type work, it is rarely found that more than 25 per cent express dissatisfaction with their jobs. Secondly, it is noticeable that even when, as in the present situation, there is considerable public attention on the effects of rationalized work, many workers—even those currently involved in industrial disputes—tend not to criticize the nature of their jobs except in terms of pay, hours, conditions, etc.[13]

This theory suggests that although workers may benefit from restructuring and organizational changes, not only may they not necessarily seek such changes, but they may even initially resist them in order to gain other benefits. Thus, whilst there may be some benefit in studying a worker's attitude to his present job in order to identify the type of changes which might be appropriate, a study of present job satisfactions will not necessarily indicate whether restructuring is required. Such a view has also been advanced by Paul *et al.*,[9] reporting work undertaken in ICI, who in effect suggest that the only way to determine whether job enrichment is appropriate is to try it. One is therefore led to the conclusion that the absence of pressures for work changes from workers does not imply that such changes would not prove beneficial. Thus whilst dysfunctional and counter-productive worker behaviour, such as high absenteeism and turnover, low quality and high grievance activity might be seen as an indication of the need for changes, the absence of these 'indicators' cannot be taken as proof of the inappropriateness of these same changes. This conclusion is of some importance in comparing experience in Scandinavia and the U.K., particularly in the motor industry. We have seen how many of the job and work changes and organizational changes in Sweden were of a remedial nature, prompted by the types of behaviour cited above. The fact that the need for remedial action is not yet evident in the U.K. motor industry could therefore be taken to indicate that such a situation will never apply, that some 'breathing space' is available before the need arises, or that preventative action is now appropriate. Since the correct interpretation is by no means clear, it is fortunate that other factors are also likely to operate as sources of pressure for such changes.

Labour

An increased mobility of labour has enabled many countries to 'side-step' labour problems by recruiting workers from poorer areas and countries. This has caused social problems both inside and outside the factory[10] and it can hardly be desirable, or possible in the long run, to continue to specify jobs that the indigenous work force tends to reject. Similarly, those companies that respond to government promptings and inducements and open new plants in economically depressed regions cannot expect to maintain a disparity in conditions of work with regard to the traditional industrial areas. Although the labour market may react against women in the short run, their equal pay, increasing involvement at the work-place and in education should preclude their becoming the new 'immigrant' labour, and should further add to the cost of labour in many industries. The present trend of improvement in the national economy has already begun to be reflected in labour shortages, and increased turnover and absenteeism, all of which effectively adds further to the cost of labour, a situation which is further aggravated by generally increasing levels of pay, social security, etc.

The manufacturing industries are increasingly likely to appear to offer unattractive employment opportunities for new employees, thus the drift to the service industries will continue. The increasing difficulty of labour recruitment (at all levels) in manufacture will therefore support efforts to increase labour productivity, which inevitably will give rise to pressures for increased mechanization and automation. The use of more female and immigrant labour is unlikely, therefore, to do more in the long run than check other, more lasting, factors. On the other hand, the trend in living standards and expectations of the work force must continue, indeed this must be one of society's major goals. The survival of conventional labour-intensive production systems will thus depend on how successfully those manufacturers who wish to continue to use such systems can do so in the face of rising labour costs, which in turn must depend upon their success in increasing labour productivity or replacing manual work.

With increasing cost of labour in the industrialized world the choice may ultimately be between an overseas plant or increased automation at home. Additionally, the orientation, expectations, values and motivational bases of the working population may further change in a manner likely to affect the willingness of workers to conform to the requirements of a production system. Thus one expedient course of action is to transfer manufacture to countries where the labour force accept monetary reward as the prime motivator. Automation may prove to be the answer for some industries, but those that retain manual methods may turn to individual or group working to combat behavioural factors. Even 'satisfied' workers may need extra motivation as social-security benefits improve, work is seen as less of a virtue and leisure activities assume more significance. By no means all recent applications of group working have been in response to such pressures, although in some cases they may have been made in anticipation. Recognition of some of the operational benefits of such methods of work organization may therefore accelerate such changes.

Efforts to automate should be greatly helped by technological developments, both in machines and in product design. However, it should be understood that behavioural factors must influence future automation, since otherwise, whilst automation will reduce labour requirements, greater problems will be associated with the fewer remaining jobs than with current largely manual systems.

Demand and markets

The continuing rise in demand for mass-produced consumer products as the standard of living increases in industrial nations is perhaps the best indication that assembly lines of one form or another will still be with us in the foreseeable future. Increased demand fortifies the basic precepts of flow-line production and will enable many new products to be manufactured on high-throughput, low-flexibility systems. Thus new 'conventional' assembly lines will continue to spring from 'individual' assembly and the economic arguments for automation will grow stronger where the technology is suitable.

Prosperity has brought new as well as increased consumer demands, both for variety and quality ('consumerism'). More flexible assembly methods are called for in many cases, and where product lines are automated robots will find increasing use, whilst when quality is a factor, group working and individual assembly may be a better proposition than conventional manual flow lines.

Industrial democracy and participation

There are signs of increasing tacit acceptance of the worker's right to a voice in the decision-making processes that affect his livelihood. Already workers have taken control of ailing factories rather than be made redundant,[11] while the idea of 'industrial democracy' has taken firm root in Western Europe, with governments and political parties considering such schemes as worker representation on the board, work councils and financial control by reinvestment of profits in equities for the work force. Once such systems are firmly established it is difficult to imagine manual workers without the kind of autonomy currently reserved for managerial and supervisory staff.

Whilst traditionally in the U.K. there has been strong interest in the principles of industrial democracy, work representation and codetermination, such interest has been moderated by scepticism about the practicality and value of such schemes, heightened perhaps by several notable failures. The current U.K. union attitude to democratization processes at the shop-floor level suggests interest but not encouragement. Traditionally in the U.K. the labour movement have sought improvement in wages and conditions rather than demanding an active part in management. It has however been suggested that in common with developments in Scandinavia and the U.S.A. future negotiations will place more emphasis on non-instrumental aspects of employment. As yet there appear to be no signs of such a development; indeed the T.U.C. is known to be sceptical of concepts such as group working, which the labour movement tend to see as a threat to solidarity and a factor encouraging company unionism which is alien to U.K. union principles. However, whilst the present situation contrasts markedly with, for example, Scandinavia, where unions and employers' federations have been the principle forces behind experimentation with company and shop-floor democratization and participation processes, a change may be due, prompted perhaps by pressures from other countries, together with a reduction in the scope of traditional labour/employer bargaining, through, for example, the wider use of labour contracts and day- and time-work payment systems.

Production systems that give a measure of autonomy to an individual or work group can be seen as an attempt at remedial action (rather than merely a compensatory reaction), which is consistent with the development of increased participation. Furthermore, changes such as job enlargement and job rotation, if seen in this context, may be judged according to their 'autonomy factor', particularly if one accepts the pretext that unskilled assembly work carries no *inherent* job satisfaction. Increased automation accompanied by the reduction in manning might therefore be seen to facilitate rather than obviate efforts to increase the degree of participation at the shop-floor level. Indeed, even when such changes result in an increase in the work rate of operatives it is possible that job and organizational changes may be appropriate, whilst such changes may be seen as the only satisfactory means to compensate workers for diminished roles. Companies such as ICI, and certain petrochemical companies (e.g. the 'Fawley agreements'[12]) have brought about considerable changes affecting the degree of autonomy of 'blue-collar' workers, and such changes, which have often been part of a larger programme of change (e.g. the ICI 'weekly staff agreement'[12]), may be seen as prototypal changes for the engineering industry. Flexibility of work-hours and holidays may be one major consequence of such changes, not only because such degrees of worker freedom are consistent with the principles of participation, but also because such factors may be seen as a likely area for future union/employer negotiations.

THE FUTURE?

It is clearly impossible to predict what type of changes will occur in particular situations because of the vast complexity of factors involved. Nevertheless, a more general and speculative look at the future is perhaps appropriate.

Manual flow lines

Although may environmental and technological forces will operate against the continuation of highly paced and rationalized assembly lines, the operational advantages of sequential processing are unlikely to be sacrificed completely in future production systems. Assembly lines are likely to exist for some time to come but will have to accommodate the need for more flexibility of production and different forms of work organization.

Possibly the least flexible flow line is the 'moving-belt' rigidly paced system. Pacing is one of the major causes of system inefficiency and, reputedly, worker alienation and low morale. Unpaced lines have been shown to be more efficient provided that small buffer stocks are allowed to accumulate between each worker or group of workers. Self-pacing by workers results in less idle time and hence greater output, but the existence of buffer stocks means more work in progress. Moving-belt lines are widely used in the assembly of large and heavy items but it has been shown that, even for such products, handling systems can be devised which enable operators to pace themselves.

For two main reasons future assembly lines will probably be shorter. Firstly, the restructuring of jobs usually results in an operator being given a longer cycle to perform and, assuming the work content stays constant, fewer operators will be needed to each line.[14] Secondly, research has shown that shorter lines can be far more efficient because there are fewer stations to which any work delays can be transmitted.

Operational control procedures must also be introduced to permit high flexibility. Techniques will be developed which can, at short notice, rebalance and reschedule the system to take into account such factors as absenteeism or changes in the product mix. The only analytical method of achieving this will be by the use of on-line computer-based heuristic or simulation techniques. However, lines may be designed with sufficiently large buffers to allow operators to work at more than one station in order to compensate for absent personnel or model imbalances. This type of system is also likely to be important in the group-working situation.

The often highly sophisticated piecework payment systems used today have in many cases proved ineffective in the assembly-line situation. The trend towards the measured day-work type of payment system, which has already started in some companies, is likely to continue althought the initial costs of changeover may be high.

Job restructuring and group working

Concern with the restructuring of an individual's work will surely be overshadowed by efforts to create autonomous, responsible work groups, within the context of worker participation and the democratization of the shop floor. Work groups pervade the industrial scene; emphasis, however, will be on formal functional groups, which are of particular relevance in respect of assembly-system design. The creation of formal work groups will provide increased opportunity for the provision of those job attributes which are commonly considered to be desirable by providing a vehicle for job restructuring. Further, unlike the restructuring of individual jobs, the creation of such groups permits the accommodation of workers who do not require restructured jobs.

A significant degree of group autonomy is unlikely to be achieved without consideration of the production process. For this reason it is likely that in certain circumstances rationalized and paced flow-line systems may be replaced by systems which permit responsible autonomous group working. Equally, however, it is clear that some of the benefits of group working will be achieved in situations which necessitate intensive material flow, low work in progress and division of operation, e.g. in the assembly of large items such as motor vehicles. It is likely, therefore, that increased attention will be given to the formation of formal functional work groups *within* flow-line systems. In fact, it is probable that in such situations the creation of formal functional work groups will be found to offer greater benefits than simple job changes through increases in cycle time, or formal job rotation. The use of comparatively highly mechanized and integrated assembly systems, e.g. automated assembly systems, reduces the potential for added work responsibilities. However, in such cases the creation of vertical work groups integrating manual workers, supervision, technical and service workers, provides scope for increased autonomy. The formation of consultative groups is most prevalent and appropriate in this form of production.

Prospects for mechanized assembly

The incentive to mechanize assembly will increase, since labour costs will represent an increasingly substantial proportion of total costs. Furthermore, technological developments in other areas of manufacturing technology will continue at a faster rate, thus assembly will represent a rising proportion of manufacturing cost.

Mechanization is evolutionary by nature and the work skills required are transitional. It is clear that many recent developments depending on partial mechanization have resulted in surviving manual jobs retaining repetitiveness, work-pacing, isolation and other negative behavioural characteristics. Future automation, particularly in view of the possible increase in emphasis on flexibility, will take account of the nature of the manual work requirements, hence the nature of manual work on assembly systems will increasingly take the form of machine-minding and monitoring.

Where industrial robots have been used to replace workers on hazardous, dirty or monotonous jobs the response has been favourable.[14] The highly automated assembly line (which included robots) at Lordstown, Ohio, did run into serious labour problems, but this may have been an effect of the negative aspects of the remaining 'in-line' manual jobs rather than a direct result of the introduction of robots. In one Swedish company[14] robots have been introduced on non-mechanical jobs in order that workers might be removed from the extremely short cycles and reduced working-space conditions of mechanized lines. The advantages of increased reliability resulting from complete automation are clear, even if for the present, partial automation continues to characterize the development of production technology, and development towards comprehensive, integrated automated systems depends on the extent to which machines can be programmed to deal with variability.

One factor which has not been examined in this study relates to the *distribution of production*. It can be argued that one means to overcome some of the problems often associated with labour-intensive mass-production systems involves the transfer of such activities to other countries, particularly the industrially developing countries, which may offer relatively cheap and plentiful labour from a work force who for various reasons are unlikely to react against rationalized and constrained manual work. The prime motivation here, at least for companies involved in 'assembly', has been expansion into new markets by horizontal integration, usually within another industrialized nation(s). As exports increase, savings in transportation and customs duties can offset the cost of setting up a new foreign plant. In addition, a price reduction and/or 'home' image can greatly influence sales, and in many cases may be necessary before the company can compete successfully in a particular market, especially if the product requires after-sales service. Occasionally, tariff barriers or import restrictions may necessitate local manufacture if markets are to be penetrated.

Assembly work is particularly suited to international transfer, especially with the advent of containerization, and may give many of the advantages of local manufacture without requiring the transfer of all managerial and technical skills and supporting industries. It can also be used as an intermediate stage, while the operational difficulties of full manufacture and the potential market value of the host country are assessed. The ability to export manufacturing technology has led to a newer phenomenon—assembly/manufacture in countries without an appreciable market but with substantially lower labour costs, so that virtually all components are shipped in, assembled and reexported to the parent company. To a great extent this has been promoted by the need to meet foreign, particularly Japanese, competition and fostered by inducements offered from governments anxious to encourage foreign investment. There is little doubt that the assembly line has suffered from Third-World (and Japanese) competition, some companies (e.g. in motor-cycle manufacture) being severely affected, whilst others have resorted to importing all their products.

Despite the restrictions that governments may inflict on foreign investment, it would seem that the potential markets of the Third World will be too tempting to ignore. Support is already growing for the idea of a 'fade-out' policy whereby after a given period ownership would pass into local hands,[15] and this would seem an excellent solution to both the problems of redressing the balance of wealth in the world and using the world's resources in a more rational manner, without provoking any immediate crises in established industries. In conclusion, the transfer of labour-intensive industry to the Third World can be seen not only as a solution to manufacturing problems but as an opportunity to alleviate the West of some of its greatest social/environmental problems, while allowing less-developed nations to improve their economies and profit from our previous experiences.

DISCUSSION

Attempts to predict the future are fraught with obvious difficulties. Nevertheless, certain fairly clear points emerge from the foregoing comments.

We can surely expect further automation and mechanization of systems, but perhaps with increased emphasis on system flexibility and hopefully with some consideration of the behavioural implications of the manual jobs required by such systems. This latter 'requirement' could encourage us to take a less pessimistic view of the nature of the recent technological changes in assembly mass production, since it can be argued that these developments, in largely ignoring behavioural factors, have been inadequate, and therefore this trend should not be extrapolated.

The fact that workers and unions in the U.K. currently exhibit somewhat different orientations to their work than their colleagues in other countries cannot surely be taken as an indication of the absence of a need to consider job restructuring and work reorganization. Certainly the U.K. is not currently subject to the same 'pressures for such changes' as for example Sweden or Italy, nor would it be quite so easy to introduce these changes in the U.K. Furthermore, the current interest in 'behaviourally oriented' changes in the U.K., if based on somewhat distorted views of the nature of the changes introduced in other countries, may be seen as a reaction rather than an expression of a need. However the situation will surely change, and the need for job restructuring and work reorganization will increase. This need should, however, be seen in the light of three factors.

(1) Doubtless there will continue to exist some very good reasons for the widespread adoption of the basic types of mass-production system identified in Chapter 2. The inherent benefits of systems such as flow lines will not be lightly abandoned, hence the need is for the development of jobs and work organization within these systems, or modified versions of the basic systems.

(2) Such modification of mass-production systems should not, however, be seen solely as a device to facilitate behaviourally oriented changes, since, as this study has shown, the modification of the parameters of systems, may have significant operational benefits. It is often argued that the system changes involved in job restructuring and work reorganization are justified only for behavioural reasons, and thus can only be justified when the indirect behavioural savings, e.g. reduction in turnover and absence, offset the necessary direct costs of reduced operational efficiency. Were this necessarily the case it would be extremely difficult to justify such changes in many situations in the U.K. at the present time. Thus, in comparison for example with the introduction of 'cell' organization in batch production, group working in mass production is often seen as a more risky and less easily justified change. Cell production can be shown to be beneficial in virtually all respects, yet it is not always clear that group working in mass production provides anything other than 'intangible' behaviour benefits. Furthermore, it is perhaps worth noting that, even with the obvious advantages of cell working, such changes have been only slowly adopted in the manufacturing industries in the U.K.

146

(3) It is unlikely that job and work changes alone will prove to be an adequate approach for the development of jobs within systems. Certainly it is likely that mere task changes, resulting in job enlargement, whilst possibly initially resulting in favourable worker reaction are unlikely to provide sustained benefits. Repetitive unskilled jobs in mass production are possibly inherently incapable of evoking the attitudes required of workers, hence manipulation of the parameters of such tasks, e.g. increased cycle time, is possibly an inadequate strategy for effective change.

Finally, we must recognize that there is surely a need to take a comprehensive approach to the development of production systems, and development within these systems. Techniques alone are unlikely to prove effective, as each situation must be treated on its own merits, although an outline policy for design, development and change may prove to be useful if not applied too rigidly. Such a policy is proposed in the following chapter.

References

1. Wilson, N. A. B., *On the Quality of Working Life*, Department of Employment Manpower Paper No. 7, 1973.
2. E.E.C. Consultative Document on Social Policy.
3. Davis, L. E., *et al.*, 'Current job design criteria', *Journal of Industrial Engineering*, **6**, 2 (1955).
4. *Selection of Assembly Systems*, IIT Research Institute, Chicago, 1971.
5. Herzberg, F., *Work and the Nature of Man*, Staples, 1966.
6. Goldthorpe, J. H., *et al.*, *The Affluent Worker: Industrial Attitudes and Behaviour*, Cambridge University Press, 1968.
7. Daniel, W. W., 'What interests a worker?', *New Society* (23 March 1972), pp. 5–8.
8. Wedderburn, D., 'What determines shop floor behaviour?', *New Society* (20 July 1972), pp. 128–130.
9. Paul, W. J., and Robertson, K. B., *Job Enrichment and Employee Motivation*, Gower, 1970.
10. Randal, J., 'French immigrant workers demand a better deal', *The Guardian* (2 April 1973).
11. Campbell, A., 'Workers across the board', *The Guardian* (20 March 1973).
12. 'ICI breaks its bottleneck', *Business Week* (9 September 1972), p. 119.
13. Sayle, M., 'Rover: where shop stewards help plan a £50m. plant', *Sunday Times* (8 July 1973), p. 8.
14. Wilsher, P., 'Fiat: where they hope £50m. will buy happy workers', *Sunday Times* (8 July 1973), p. 9.

CHAPTER 12

A Design/Development Policy

The objective in this concluding chapter is to attempt to piece together the views that have been stated and to complete the 'picture' that has begun to emerge in earlier chapters. Certain principles will be identified or restated in order to permit the identification of a design/developmental policy of the type advocated in the previous chapter. The principal focus will be the nature and organization of the jobs associated with systems—this subject being seen as one of major future importance. It is hoped to be able to show that a common basic design/developmental policy might be appropriate irrespective of the nature of the systems used, although it is accepted that the technology employed will constrain the application of this policy.

The identification of these principles, and the development of the policy, is founded upon the three premises generated in the previous chapter, i.e.

(1) That the provision of restructured and enriched manual jobs is compatible with the use of the basic systems of mass production identified in Chapter 2.
(2) That the continued utilization of these basic systems of mass production is both economically and technically desirable.
(3) That the desire to provide enriched jobs is an appropriate objective in the light of possible future pressures and requirements.

DEVELOPMENT PRINCIPLES

Basic observations

The information presented in Chapter 5, together with the observation of Chapter 6 permit the statement of the following basic observations.

(a) That what is often referred to as job restructuring can be identified as the introduction of

  (i) Job and work changes providing either job enlargement (a horizontal change) or enrichment (a vertical change).
  (ii) Organizational change, primarily job rotation or self-organization, both of which facilitate or in some circumstances are prerequisite for job enlargement and enrichment.

147

148

(b) That an examination of the means by which jobs might be restructured requires the identification of certain sets of characteristics (tasks, task relationships, work method, organization and opportunities), manipulation of which might lead to the provision of jobs having desirable attributes.

(c) That the creation of formal functional work groups is the principal form of organizational change and that the nature, and benefits of this form of work organization, are largely dependent upon the extent of the autonomy and responsibility assigned to the group.

(d) That organizational change, the provision of group working and the effective restructuring of jobs requires the increased participation of workers in planning and control activities and the 'deverticalization' of organizational hierarchies (i.e. the delegation of autonomy and responsibility to lower organizational levels).

Evaluation of (a) and (b) leads to the formulation of the model shown in Exhibit 12.1, which combines the two exhibits from Chapter 5. The figure

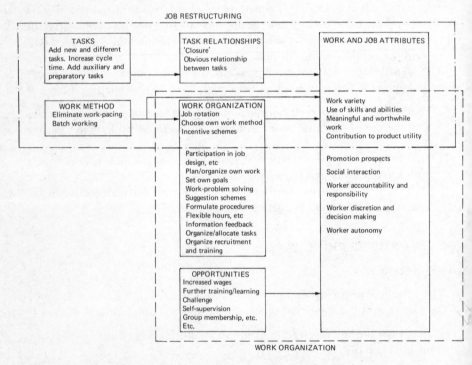

Exhibit 12.1 The nature and means of job restructuring

shows how the manipulation of job characteristics provides for the creation of jobs having some of the attributes which are commonly held to be desirable. The distinction between job restructuring and work organization was introduced in Chapter 5, the importance of the latter deriving from the fact that the simple manipulation of tasks and work methods can lead only to the provision of a limited number of desirable attributes. Such changes are largely horizontal, and with organizational changes, which facilitate vertical job and work changes, many of the more important job attributes are unavailable. Furthermore, Exhibit 12.1 suggests that organizational change, without concomitant changes in tasks, task relationship and work method, *can* provide most if not all of these job attributes. Thus work organizational change, particularly self-organization, must be an important part of any design/development policy whether jobs can or cannot be enlarged, and a necessary part when task and work-method changes give rise to further work rationalization (e.g. in assembly mechanization). Points (c) and (d) suggest some objectives of, as well as certain means for, work organizational change. These objectives relate primarily to the formation of autonomous and responsible functional work groups, whilst a means for the achievement of this objective is the delegation of some responsibilities from other personnel. Such *deverticalization* affects both workers and those normally associated with them, i.e. supervisory and certain technical and auxiliary staff, and provides a powerful means for job enrichment.

A DEVELOPMENT POLICY

The above discussion suggests two important principles or dimensions for a development policy, whilst indicating the end result and manner of their application. These dimensions are outlined below and in Exhibits 12.2 and 12.3. Job design is here considered to cover job restructuring as identified in Exhibit 12.1, whilst work organization covers the organizational changes also dealt with above.

(1) *Job design* (Exhibit 12.2). Following the objectives outlined in Exhibit 12.1 it is proposed that workers' jobs should provide 'closure' and an 'obvious relationship between tasks'. This requires the performance of tasks which might otherwise have formed subsequent and prior operations. Individual job design is completed by consideration of work methods, in particular the minimization of the pacing effect and constraints provided by the production system.

(2) *Work organization* (Exhibit 12.3). 'Deverticalization' of the organization should increase the responsibility of workers, enhance job attributes and possibly improve system flexibility and communication. The formal functional grouping of related jobs (i.e. workers associated with the same product) as an essentially self-contained, autonomous unit provides the final, probably most important, aspect of work organization. Depending upon the extent of the delegation of tasks to workers, such small integrated productive groups may contain certain indirect jobs and should provide the benefits of group working identified in Chapters 7 and 8.

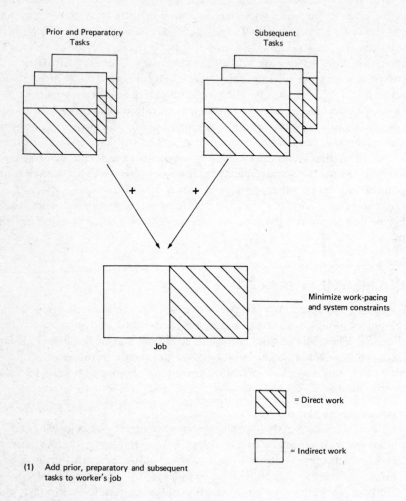

Exhibit 12.2 Suggested principles of 'job design'

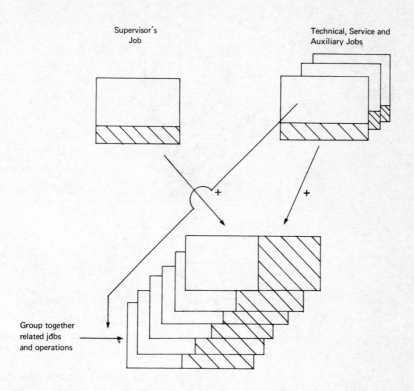

Supervisor's
Job

Technical, Service and
Auxiliary Jobs

Group together
related jobs
and operations

(1)   Delegate aspects of supervisors,
       technical, service and auxiliary jobs

(2)   Group together related jobs

Exhibit 12.3 Suggested principles of 'work organization'

These two principles are combined in Exhibit 12.4 where the third 'principle' of organizational design is introduced in <u>recognition of the fact that neither workers nor production systems exist in isolation in a plant, but as part of a</u>

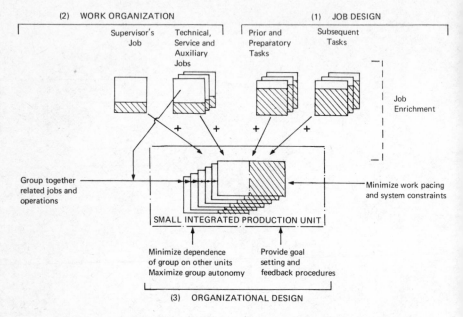

Exhibit 12.4 Principles for the design and development of systems and jobs

<u>larger organization.</u> Thus the third principle or dimension of our proposed policy for system and job design and development is as follows.

(3) '*Organizational design*' deals with the environment for the production unit, i.e. the provision of information systems for goal setting, feedback and performance measurement, together with the necessary support systems which ensure the viability and continued existence of the unit.

Discussion

The model shown in Exhibit 12.4 depicts the proposed policy for system and job design and development, based on the application of three principles. Whilst these three simple principles are, *per se*, in no way novel or original, the proposal of a policy based on the application of all three principles does imply an approach which contrasts somewhat with much of what is now considered to be standard practice. For example, job enrichment—often advocated as the solution to problems of repetitive rationalized work—is seen to involve only two of these three principles (Exhibit 12.4), whilst job design is here viewed as one, perhaps minor, dimension dealing only with direct productive work.

The cases discussed in Chapter 6 indicated the desirability of considering work organization (particularly group working) in job and in system design; however, in very few cases had this approach been taken to the extent proposed above, i.e. the creation of small, integrated and essentially self-contained production units. The creation of such organizational units is here seen to be not merely desirable but an important part of a comprehensive design/development policy. This clustering of jobs, together with the deverticalization of the organization, is seen to be a powerful device from both organizational and behavioural viewpoints. The need to provide for the existence of a unit within a unit within a department or plant necessitates some consideration of the environment in which the unit is to operate. The effect of job restructuring and work-organization changes on plant and departmental organization have been observed in many of the study visits, but in few cases were such factors seen as an integral part of the design or development process. The novelty of the above policy is therefore seen to be the comprehensiveness of the approach and the nature of the end result rather than the originality of the fundamental principles.

Thus any originality in our approach derives from its scope rather than the nature of its component parts. Such scope would, in practice, necessitate the involvement and contribution of many functions and specialists in an area for which we currently have no appropriate name. Our approach is wider than that conventionally implied by job design or job restructuring, might be seen as an aspect of organizational development, yet also relates to production or operating system design.

DEVELOPMENT PROCEDURES

System constraints

The manner of the application of these principles may be influenced by the nature of the production system employed. For this reason no attempt is made to specify detailed 'tactics' for development within systems, but rather an outline of an appropriate policy. The nature and characteristics of the system impose greatest constraints on work organization, yet in all but the highly paced, rationalized and machine-controlled systems, some degree of formal functional group working is possible. Furthermore, work reorganization is perhaps the only effective means for job restructuring where system constraints on the individual are high.

The three-part approach presented above permits the maximum exploitation of the freedom available 'within' systems. Indeed it is perhaps because of their adoption of a narrower approach, typically emphasizing job and work changes and ignoring work organization, that many writers have concluded that provision of satisfying work necessitates the abolition of flow-line working.

The basic principles set out above are considered to provide guidelines appropriate for use in the design of new systems and development of existing systems of each of the types identified in Chapter 2 and for each of the possible methods of operation. Technological change, particularly increased mechanization and automation, may therefore be seen as presenting an opportunity for the application of these principles rather than, as has often been the case, a further immutable constraint. Furthermore, this policy is considered particularly appropriate in situations where job rationalization changes are to occur (e.g. in the establishment of Group-Technology production). In such cases the major system constraints may relate to job design, whilst work organization and organizational design provide a means for *compensatory* changes.

Participation

Exhibit 12.5 outlines the five stages that might be involved in the application of the policy; however, development within a system would exclude stage 1, although system parameters may be changed. An alternative procedure is shown in which the formation of the production unit, i.e. the grouping of jobs, precedes work organization. This procedure provides for greater worker involvement in development, since in this case work organization may be the responsibility of the work group. Although Exhibit 15.2 implies the design of *individuals'* jobs, job design may equally involve the allocation of tasks and responsibilities to a previously defined group. It can be argued that participation can only be introduced by participative methods. Such an approach is surely desirable, especially if we recognize

(1) That management will probably never be in a position to predict precisely what type of jobs and work organization are most suited to particular workers in particular circumstances.
(2) That if those affected have not contributed to the specification of changes, this alone may be sufficient to promote resistance to changes irrespective of their merits.

Equally, however, because of a possible instrumental orientation to work, workers may not actively press for changes, or seek to influence the design of new systems. Thus management must assume responsibility for initiating such participation, if only by establishing consultative procedures.

The role of management

Deverticalization will affect management no less than workers, whilst work organization and organizational change, in general, will have repercussions which may promote resistance to change. Thus management is both the means for, and target of, change. Whilst the integrated production units created through the application of this policy may require less direct supervision, their creation is likely to be a major managerial task in terms of planning, creation

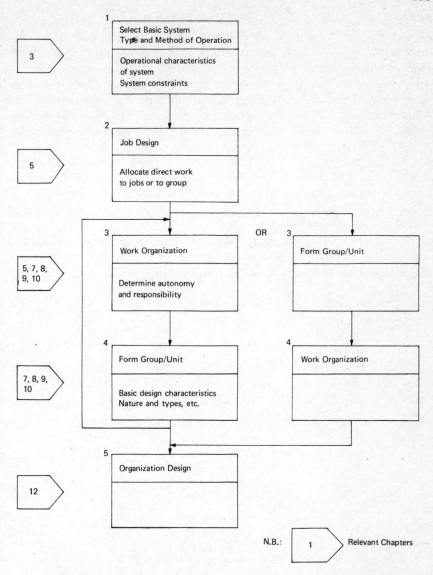

Exhibit 12.5 Procedures for design/development of systems and jobs

and maintenance. Furthermore, the creation of such units may affect 'outside' organizations such as customers and suppliers, especially when technical and auxiliary functions are integrated within units and when workers make direct contact with such organizations and individuals.

Payment, communication, training, supervisory, negotiation and promotion procedures may all be affected. Indeed the application of the three principles may be seen as having a possible catalytic or precipitating effect. Multiplier, rather than simply additive, effects may occur; furthermore, since the nature of individuals needs, attitudes and behaviour are largely unpredictable, a mechanistic approach to change is probably inappropriate. A pragmatic, rather than technique-oriented, approach is required which recognizes that each situation must be treated on its own merits; thus no two units may be alike and the speed and nature of the development of units may vary substantially. This further suggests that without adequate preparation both of and by management, and without conviction, such changes may be ineffective or impossible.

References

1. Daniel, W. W., and McIntosh, N., *The Right to Manage*, P.E.P., 1972.
2. Ford, R. N., 'Job enrichment lessons from AT & T', *Harvard Business Review* (January/February 1973), pp. 96–106.
3. Wilson, N. A. B., *On the Quality of Working Life*, Department of Employment Manpower Papers No. 7, 1973.
4. Schumacher, E. F., *Small is Beautiful*, Blond and Briggs, 1973.

APPENDIX A

Job Restructuring Exercises— Blue-collar Workers

Job restructuring exercises involving blue-collar workers

(1) Rearrangement/replacement of assembly-line work

Company	Source	Initial job	Reason	Revised job	Preparation	Payment	Result	Comment
Manufacturer of small items, U.K.	A. Wilkinson (1970)	Assembly-line work	High labour turnover	Structured work				
R. Baxendale and Sons Ltd., U.K.	L. K. Taylor (1972)	Assembly of gas convector heater on nine-stations conveyor line		Individual responsible for complete assembly				
Texas Instruments, U.S.A.	F. K. Foulkes (1969)	Women assembling very complex instrument project-oriented group	Uneconomic production	Use of problem solving goal-setting led to rearrangement of assembly procedure			Time reduced from 138 hours to 32 hours Absenteeism labour turnover, learning time, complaints, health centre visits all reduced. Improved attitudes	500 involved
IBM, Poughkeepsie, New York, U.S.A.	D. Wharton (1954)	Wiring 2331 wires on panel. One girl for black, one for green, one for red and one for violet wires		One girl on one panel				

Maytag, U.S.A.	E. Conant and M. Kilbridge (1965)	Assembly of 27-part water pump	Cost reduction and quality improvement necessary	One man assembles one pump and inspects own work	Experiments with six- and four-station lines	Annual savings of $22,000	
Cadillac Factories, Detroit, U.S.A.	G. Friedman (1961)	Manufacture of high-precision parts for aircraft engines		One woman completes assembly		Improved morale and output	Wartime
Maytag, U.S.A.	E. Conant and M. D. Kilbridge (1965)	Assembly of top cover for automatic washing machine with 46 parts		One-man assembly		Quality improved, costs reduced	
Maytag, U.S.A.	E. Conant and M. D. Kilbridge (1965)	Assembly on 100-man line of commercial washers		7 stations for progressive work on stationary product. Nearby inspection			
IBM, Lexington, Kentucky, U.S.A.	J. Anderson (1970)	Assembly-line work		One man assembling functional part of electric typewriter. Self-inspection			
Philips, Eindhoven, Holland	NV Philips Report (1969) H. van Beek (1964)	104-woman line on televisions. No buffer stocks. Final inspection	Persistent problems	Test line with groupings of 29, 28, 14, 17 and 16 women. Stocks and inspectors for each group		Output and quality improved. Girls preferred smaller groups	

(1)—continued

Company	Source	Initial job	Reason	Revised job	Preparation	Payment	Result	Comment
	A. Marks (1954)	29 girls on line making hospital appliance. Rotation between 9 stations		One girl assembling complete unit. Controls work sequence, procure supplies and inspect	Elimination of conveyor. 9 stations each with buffer stocks		Elimination of conveyor led to differences in output and improved quality. Individual working increased output and quality	
Corning Glass works, Medfield, U.S.A.	J. Gooding (1970)	Assembly of digital electrometers		Girl doing complete assembly		Under review	Profits up, inspection and rejects down. Fewer inspectors required	Complaints about lack of extra payment
Non-linear Systems, Inc., California, U.S.A.	A. Kuriloff (1963)	Manufacture of instruments	Test of McGregor's 'Theory Y'	7-man, self-paced, work teams. Responsible for supply, testing, quality and training. Organized on whole task basis			30 per cent improvement in man-hours per instrument and improved quality. 70 per cent fewer complaints about the products	Introduced too quickly
Philips, Hamilton, Scotland	A. Leigh (1969)	Production of domestic appliances		4 or 5 operators in groups producing complete subunits			10 per cent increase in output, 50 per cent reduction in defects, improved job satisfaction	

Philips	A. Leigh (1969)	Coil winding into 4 separate areas and sub-divided into short-cycle, simple jobs	High turnover, amongst those capable of more skilled work	All operations grouped into one job and carried out on specially constructed tables	Training	30 per cent saving in labour costs, doubling of quality, less stock in transit, less transport, training costs recovered in 12 months	Less popular with young workers. Management structure altered
Philips	N.V. Philips Report (1969)	Assembly line of 11 or 22 people working on electrical subassemblies. Cycle time 50 seconds		One person initially tried for complete assembly but found to require groups of 3. 13-minute cycle	Improved lay-out planning	Output increase of 3 per cent plus improved quality. Turnover and absenteeism reduced	Less popular with young workers.
Philips Ind. Pty. Ltd., Sydney, Australia	T. P. Pauling (1968)	Line producing either (a) subassemblies or (b) completed radios. Batch production	Technical and administrative problems	One operator doing complete assembly from basic components. Inspection. Interpretation of drawings	None except during training	Stock saving $300,000 Elimination of subassembly obsolescence. Fewer faults. Mixed feelings	10 workers involved
Producer of drive mechanisms, Iowa, U.S.A.	G. Tuggle (1969)	Carousel-type assembly line with 8 stations		Modular assembly system with 5 stations, where complete job is performed and unit tested. 6-minute cycle time	Materials handling system needed. Complete redesign	37 per cent reduction in labour costs. Reject rate down from 5 per cent to 0·5 per cent	

(1)—*continued*

Company	Source	Initial job	Reason	Revised job	Preparation	Payment	Result	Comment
Manufacturer of domestic furnaces, Wisconsin, U.S.A.	G. Tuggle (1969)	Batch production on assembly line with runs of 2 or 5 days—2 lines, 5 products	Customer complaints about product quality	8 independent assembly lines each with 2 workers. Cycle time 10 to 12 minutes quality test			80 per cent reduction in complaints over 3 months. Gain in scheduling, flexibility and lower finished goods stocks	
Appliance manufacturer, Wisconsin, U.S.A.	G. Tuggle (1969)	5 product model. 2 conveyors, 1 minute cycle time. 1 line on electrical equip, 1 line on hardware	Inventory problem	Man–women team assembles whole unit	Materials handling system redesigned		24 per cent reduction in assembly labour costs. Weekly changes in schedules permit optimum product mix in inventory	
Wireless manufacturer, U.K.	G. Friedman (1961) D. Cox and K. M. D. Sharp (1951)	10 workers on assembly line with 2-minute cycle time		Woman making unit. 20-minute cycle time			Initially output fell from 10th to 23rd week 18 per cent better than line. Better quality	
Wireless manufacturer, U.K.	G. Friedman (1961) D. Harding (1931)	Assembly line	Attempt to give meaning to job	10 wires instead of 2 to be soldered by worker			Output and job satisfaction increased	

	W. Lawless (1971)	Conveyor belts		operator makes complete subassembly	selection of personnel	
Philips Ltd., U.K.	T. Kempner and R. Wild (1973)	Assembly of electro-mechanical component—flow line, female workers	Experiment	Individual assembly and inspection		No general improvement in output or quality
Company X, U.S.A.	D. Sirota (1973)	Assembly of office equipment. 17 operations on long flow line. Cycle time 4 minutes		Create short lines with larger jobs		Quality improved. Easier to identify sources of defects
Electrical equipment manufacturers, U.S.A.	E. F. Huse and M. Beer (1971)	Flow-line assembly of hotplates—female workers	O.D. program	Individual assembly to complete plate. Eventually given responsibility for final inspection		Fall in controllable rejects from 23 per cent to 1 per cent in 6 months. Absenteeism down 8 per cent to 1 per cent in 6 months. Increase in productivity of 84 per cent.
Electrical equipment manufacturers, U.S.A.		Instrument assembly. Female workers		Individual assembly		Easier introduction of new product. Improvements: production (17 per cent), quality (50 per cent) and absenteeism (50 per cent)

(1)—*continued*

Company	Source	Initial job	Reason	Revised job	Preparation	Payment	Result	Comment
Philips	H. J. van Beek (1964) Philips Report (1969)	104-station line		Reduced effective line length by division into 5 groups				Reduced waiting time. Reduced absenteeism. Quality improved. Reduced system loss
Pharmaceutical Appliances, U.S.A.	L. E. Davis and R. Canter (1956)	Flow line. 29 women workers. Job rotation	Experiences	9-station lines and job rotation			Productivity down. Quality up	
Pye Ltd., Lowestoft, U.K.	*Mfg. Mgt.* (Dec. 1972)	Flow-line assembly of televisions	Increase variety. Improve quality	Teams of 7 or 8 workers. Responsibility for quality			Reduced turnover. More flexibility	
Volvo, Sweden	P. McGarvey (1973)	Automobile assembly	New plant	Groups of workers perform more work on cars without machine pacing				
Philips	Philips Co. Report (1969)	Assembly lines of 22 or 11 workers. Cycle time 50 seconds		Assembly of item by group of 3 workers			Output up. Quality improved	
IBM, Amsterdam	N. Foy (1973)	Assembly-line work females. 3 minute cycle time	Output changes. Quality and large turnover problems	Used shorter mini-lines with longer cycle times			Productivity up. Turnover down	

Company	Source	Initial job	Reason	Revised job	Preparation	Payment	Result	Comment
Meccano Ltd., U.K.	D. Harvey (1973)	Moving belt flow-line. Women workers	To increase productivity	Individual assembly	4-month induction period	New individual incentive system introduced	Productivity up. Flexibility increased. Total time required reduced from 1·19 minutes to 0·675 minutes	

(2) Workers given additional responsibility.

Company	Source	Initial job	Reason	Revised job	Preparation	Payment	Result	Comment
Plumbing firm, Sweden	A. Wilkinson (1970)	Service men	Too costly and inefficient	Increased responsibilities		New incentive scheme		
Imperial Chemical Industries, U.K.	W. Paul and K. B. Robertson (1970)	Toolsetters	Need for high quality	Given responsibility for efficiency of certain equipment			Sustained improvement in output	4 setters and 28 workers involved
IBM, U.S.A.	J. Maher and Woverbach (in Maher (1971))	Inspection of both manufactured and bought parts	Poor quality and low morale	Machine quality analysis given much wider responsibility, as were inspectors, in dealing with other people			Improve morale, and acceptance rate up from 92·9 per cent to 97·5 per cent. Fewer inspectors	

(2)—*continued*

Company	Source	Initial job	Reason	Revised job	Preparation	Payment	Result	Comment
R. Baxendale and Sons Ltd., U.K.	L. H. Taylor (1972)			Individual works out his own goals feedback and recognition are given also involved in improving methods				
William Tatton and Co. Ltd., U.K.	L. K. Taylor (1972	Wrapping Department		Increased responsibility to deal directly with service departments and attend production meetings				
Dow Chemical Co., U.S.A.	L. Davis and R. Werling (1960)	Distribution and maintenance departments each had maintenance crew for 60 per cent to 75 per cent of its needs	Need for improved productivity and no trade demarcations	When centralized maintenance departments formed, skill and functional changes made to jobs of other workers	Formal 'on' and 'off-the-job' training	Job reclassification and wage increase	Reduction in cost over 2 years from 30 to 110. Quantity and quality improved. Labour cost down 90 to 65 in 2 years	
Aircraft instrument manufacturer, U.S.A.	L. Davis and E. Valfer (1965)			Product responsibility and quality responsibility given via supervisor to the group			Quality and cost improvement. Improved attitude towards supervision	

Organisation	Source	Job	Reason	Change	Method	Results	Payment
Jensen Motor Co., U.K.	L. K. Taylor (1972)	Assembly-line workers		Inspectors eliminated		Saving of 30 inspectors. 50 per cent reduction in rework and scrap over 1st year. Greater interest in job	P.P.R. with time allowed for inspecting
Graflex, U.S.A.	E. J. Tangerman (1953)			Operators inspect parts and are held responsible		Improved attitudes equal performance	
Company X, U.S.A.	D. Sirota and A. D. Wolfson (1972)	Final assembly of electronic devices	Concern for morale of employees	Assemblers given responsibility for testing the devices	Diagnosis; top management exposure; training		
Company X, U.S.A.	D. Sirota and A. D. Wolfson (1972)	Machine tending of slicing equipment. Rotation between machines	To improve employee morale	Each operator made responsible for 2 machines, their maintenance and blade replacement	Training performance feedback system developed	Improved performance, reduced maintenance costs, improved employee attitudes	
Company X, U.S.A.	D. Sirota and A. D. Wolfson (1972)	Assembly of electronic devices. 50 minute cycle time.		Five operations combined to one job. 4-hour cycle time		25 per cent reduction in building-up time. Increased flexibility preferred by workers	
Swedish sack manufacturer	A. Wilkinson (1970)	Machine minding	Increased manning led to slower productivity	Workers given responsibility to make good rejects and decide output levels		Improvement in output and reduction in wastage	New payment system

(3) Rotation of jobs

Company	Source	Initial job	Reason	Revised job	Preparation	Payment	Result	Comment
Sun Oil Co., U.S.A.	C. Argyris (1959)		New plant	All operators trained to do all the different kinds of work				
Polaroid, U.S.A.	F. K. Foulkes (1969)	Production workers with routine jobs	Pathfinder programme to make jobs more interesting	Half time spent in a more challenging job such as R. and D., the Library, Office or sales	Training and 3-month trial	No loss of earnings	53 moved permanently to other job. Old job became a source of dissatisfaction	Desire for better social status caused some social problems
	F. K. Foulkes (1969)	Assembly work		Employees given responsibility for carrying out high-content inspection, job rotation				
Philips	N.V. Philips Report (1969)	15 men on assembly work on 7 different jobs in exacting conditions, needing relief every 2½ hours		Job rotation every 10 minutes	Inadequate preparation		Gain of 10 per cent on man-hours used for light setting-up and maintenance as well as administration. Abandoned after 1 year	Wastage of material increased. Charge-hand did the better jobs
Polaroid, U.S.A.	F. K. Foulkes (1969)	Production workers with routine jobs		Employee given 6-month exposure period in more challenging job then allowed to apply for full-time change				

Company	Source	Revised job	Payment	Comment
Hoover, U.K.	Management Today (May 1973)	Work teams with job rotation		Approximately 10 per cent of total work force
Saab-Scania, Sweden	D. Palmer (1972)	Job rotation		
U.S.A.?	H. Rosen (1963)	Rotation daily		Increased job satisfaction

(4) Workers responsible for additional and different types of task

Company	Source	Initial job	Reason	Revised job	Preparation	Payment	Result	Comment
IBM, Endicott, U.S.A.	C. Walker (1950)	Jobs of machine operator, set-up, man, tool-sharpener and inspector		Incorporated into one job; determined output norm with foreman		Higher wages	Improved quality, less idle time, status of worker improved	
William Tatton and Co. Ltd., U.K.	L. K. Taylor (1972)	Steaming and coating department		Increase in job skills and range of tasks allocated to individual workers				
IBM's French and other European factories	D. Wharton (1954)	Machine operators		Carry out own set-up				

(4)—continued

Company	Source	Initial job	Reason	Revised job	Preparation	Payment	Result	Comment
Philips	N.V. Philips Report (1969)	Machine operators	Economic reasons	Maintenance where possible quality inspection	Initially supervisors removed training		Production at former level following reintroduction of supervisors	
Power Plants, U.S.A.	F. Mann and R. Hoffman (1960)	Process workers in 3 categories (1) Boiler, (2) Turbine and condenser, (3) Electrical; status division within groups	Construction of new plant	One class of operator with 3 skill classifications (1) Operator A, (2) Operator B, (3) Helper. Rotation within classifications			High level of satisfaction in enriched jobs	

(5) Control of work speed

Company	Source	Initial job	Reason	Revised job	Preparation	Payment	Result	Comment
Hovey and Beard, U.S.A.	W. F. Whyte (1955)	Women spraying wooden toys and replacing them on moving hooks	absenteeism and turnover high; turnover low	Women given control over speed of belt—8 involved	Meetings to discuss problems		Increased productivity without reduction in quality	High earnings compared with rest of plant led to trouble

(6) Self-organization

Company	Source	Initial job	Reason	Revised job	Preparation	Payment	Result	Comment
Electric and electronics firm, Denmark	A. Wilkinson (1970)		Poor personal relations in work groups	Group decides own make-up				

					Training of operatives	Changes made		
Building materials firm, Sweden	A. Wilkinson (1970)			Groups organize own shift system multi-skilled men. Interchangeable			Resistance met from some supervisors	
Barry Corporation, U.S.A.	J. Gooding (1970)			Introduction of work teams				
American Telephone and Telegraph Co. Ltd., U.S.A.	R. N. Ford (1969)	High turnover	One team writes up order, second makes the connection, a third tests the circuit; manager receives unit index on errors	One team organized and does complete job	Meetings to produce ideas 'Greenlight sessions'		Greater productivity fewer faults, fewer personnel and fewer union grievances	40 staff involved
Volkswagen Motors Ltd., Germany	L. K. Taylor (1972)	Assembly-line work		Group held responsible for quality. Participation in decision-making on whole work on subassemblies. Requisition own materials and equipment. Group budgeting. Authority to discard faulty materials, draw up schedules, initiate investigation				Throughout factory

(6)—*continued*

Company	Source	Initial job	Reason	Revised job	Preparation	Payment	Result	Comment
Non-Liner Systems Inc., U.S.A.	A. H. Kuriloff (1963)	Assembly lines. Male workers		7-man teams build complete item. Team allocation of tasks			Production increase	
Electrical equipment manufacturer, U.S.A.	D. Sirota and A. Wolfson (1972)	Assembly of power supplies on flow line		3–5-man teams. Own quality audit and work allocation			Quality and quantity improved. Improved flexibility, lower absenteeism	
Harwood Manufacturing Co., U.S.A.	N. Viteles (1953)	Clothing manufacturer. Group of female workers	Resistance to change	Experiment in group participation in decisions relating to job changes				Output increase
Harwood Manufacturing Co.	N. Viteles (1953)	Clothing manufacturer. Groups of 1–12 sewing machine operators		Teams asked to set own production goals				Output increase
Electrical equipment manufacturer, U.S.A.	E. F. Huse and M. Beer	Forming tubes on lathes—females	O.D. programme	Teams of workers responsible for total task. Responsible for schedules and allocation of work				Eventually improved output and worker commitment and involvement

Company	Reference	Type of work	Problem	Changes	Results
Hellerman Deutsch (Bowthorpe Group)	L. K. Taylor (1972)	Manufacture of high-precision components		Participation in making improvements. Training to make worker more flexible. High degree of self-supervision	8 groups
G.E.C., U.S.A.	M. Sorcher and H. Meyer (1968)	Assembly workers		Comparison of high responsibility discretion, variety and freedom of movement with usual groups	
Philips	N.V. Philips Report	3 foremen, 15 chargehands, 231 assembly workers and inspectors 5 to 120 second cycles		Formation of independent groups responsible for allocation, supply of materials, quality inspection within the group. Representation at departmental talks	Due to faster feedback of faults wastage was reduced from 7 per cent to 3 per cent. Greater job satisfaction
Saab-Scania, Soedertaelde, Sweden	D. Palmer (1972)	Assembly-line making engines at rate of 30 per hour	Labour turnover 100 per cent p.a. and high absenteeism	7 work areas with 4 fitters in each to form a team each assembling whole engine comprising 90 parts. Group organize own work—30 minutes/engine	Cost approximately £3000 per annum for space and tools

(6)—continued

Company	Source	Initial job	Reason	Revised job	Preparation	Payment	Result	Comment
Volvo, Gothenberg, Sweden	D. Palmer (1972)	Upholstery department assembly workers		Participation in decision-making and elimination of foremen				
G.E.C., U.S.A.	Business Week (9 Sept., 1972)	Welders		Formation of groups responsible for scheduling and planning work load			Greater commitment to job. Quality and efficiency improved	
General Foods, Topeka, U.S.A.	Business Week (9 Sept., 1972)	Production workers		Assignment of tasks to teams of 7 to 17 members. Workers learn each job. Each team covers entire phase of operation from processing raw material to end product, packaging shipping and office work				
T.R.W., U.S.A.	Business Week (9 Sept., 1972)	Assembly-line work		Group assembles product scheduling their own work				
Company B., U.S.A.	W. Skinner (1971)	General-purpose job shop	Low labour production	Set own rules and regulations to achieve required output. Check quality schedule	'Trainer' available to help			30 workers involved

Company	Author	Nature of work	Experiment	Description	Results
Donnelly Mirrors, U.S.A.	J. F. Donnelly (1971) J. Gooding (1970)			Interlocking work teams, planning and problems solving by team	No difference in output. Thought to be
Footwear manufacturer, Norway	J. R. P. French, et al. (1950)	Assembly of footwear by group of female workers (makes and few)	Experiment	Control groups and experimental groups. Experimental groups given participation in introduction of new models	
Texas Instruments, U.S.A.	F. K. Foulkes (1969)	Production work. Capital intensive operation		Employees given cost information to help in problem solving	Plant tours to see group contribution
H. P. Hood and Sons, Boston, U.S.A.	F. K. Foulkes (1969)			Watched movie of work and suggested new layout to improve efficiency	Saving of $10,000 for an outlay of $500 Group of 2 men and 8 women
Saab-Scania, Sweden	D. Palmer (1972)			Discussion about work methods, safety, work environment representation at development group	
Donnelly Mirrors Inc., Holland, Michigan, U.S.A.	J. F. Donnelly (1971)			Work teams plan and organize, work and develop cost-reduction method	Saving at rate of $300,000 per annum after 1 month. Rejects down from 13 per cent to 6 per cent. Improved efficiency

(6)—continued

Company	Source	Initial job	Reason	Revised job	Preparation	Payment	Result	Comment
Motorola, U.S.A.	D. Sirota and A. Wolfson (1972) J. Powers (1972)	Machine workers		Elimination of Inspectors. Groups used for decision making. Work on complete modules				
I.B.M., Endcott, U.S.A.	D. Wharton (1954)			Workers involved in product design				
Convair Division of General Dynamics Corporation, U.S.A.	A. Kuriloff (1966)			Members of Workshop assist in decision			Cost saving	

(7) Non-categorized experiments

Company	Source	Initial job	Reason	Revised job	Preparation	Payment	Result	Comment
Pulp and Paper Mill, Norway	A. Wilkinson (1970)							
Italian subsidiary of Belgian company	A. Wilkinson (1970							

British company	A. Wilkinson (1970)
Cryovac U.S.A.	J. Powers (1972)
Precision Castparts, U.S.A.	J. Gooding (1970)
Proctor and Gamble	J. W. Anderson (1970)
Toledo plant, Chrysler, U.S.A.	J. W. Anderson (1970)

References

Anderson, J. W., 'The impact of technology on job enrichment', *Personnel*, **47**, 5 (September/October 1970), pp. 29–37.

Argyris, C., *Integrating the Individual and the Organisation*, Wiley, 1964.

Beek, H. J. van, 'The influence of assembly line organisation on output quality and morale', *Occupational Psychology*, **39** (1964), pp. 161–172.

Business Week, 'Management itself holds the key' (9 September 1972).

Conant, E. H., and Kilbridge, M. D., 'An interdisciplinary analysis of job enlargement: technology, costs and behavioural implications', *Industrial Labour Relations Review*, **XVIII** (1965), pp. 377–395.

Cox, D., and Sharp, K. M. D., 'Research on the unit of work', *Occupational Psychology* (April 1951).

Davis, L. E., and Canter, R., 'Job design research', *Journal of Industrial Engineering*, **7** (1956), p. 275.

Davis, L. E., and Wearing, R., 'Job design factors', *Occupational Psychology*, **XXIV** (1960).

Davis, L. E., and Valfer, E. S., 'Intervening responses to changes in supervisor job designs', *Occupational Psychology*, **XXXIX** (1965), pp. 171–189.

Donnelly, J. F., 'Increasing productivity by involving people in their total job', *Personnel Administrator*, **34**, 5 (September/October 1971).

Ford, R. N., *Motivation Through the Work Itself*, American Management Association, New York, 1969.

Foulkes, F. K., *Creating More Meaningful Work*, American Management Association, New York, 1969.

Foy, N., 'A blow at the factory automation', *The Times*, London (5 March 1973).

French, J. R. P., 'Field experiments; changing group productivity', in Miller, J. G. (Ed.) *Experiments in Social Process*, McGraw-Hill, 1950.

Friedman, G., *The Anatomy of Work*, Glencoe, Illinois Free Press, 1961.

Gooding, J., 'It pays to wake up to the blue collar worker', *Fortune* (September 1970).

Harding, D. W., 'A note on the subdivision of assembly work', *Journal of the National Institute of Industrial Psychology* (January 1931).

Harvey, D., 'Better ways to put the pieces together', *Business Administration* (April 1973), pp. 82–83.

Huse, E. F., and Beer, M., 'Eclectic approach to organisational development', *Harvard Business Review* (September/October 1971), pp. 103–112.

Kempner, T., and Wild, R., 'Job design research', *Journal of Management Studies*, **10**, 1 (1973), pp. 62–81.

Kuriloff, A. H., 'An experiment in management—putting theory to the test', *Personnel* (November/December 1963), pp. 3–17.

Leigh, A., 'Making work fit', *Business Management*, **99**, 8 (September 1969), pp. 46–48.

Mandle, J., and Lawless, J., 'Anxiety grows over the stress factor', *Industrial Management* (July/August 1971).

Mann, F. C., and Hoffman, R. L., *Automation and the worker*, Holt, Rinehart, Winston, 1960.

Manufacturing Management, 'Work teams at Pye beat production line problems' (December 1972), pp. 18–19.

Management Today, 'Hoover's group therapy' (May 1973).

McGarvey, P., 'Car making without the boredom', *Sunday Times* (4 February 1973).

Marks, A. R. N., Unpublished Ph.D Dissertation, Univ. of California, Berkeley, U.S.A. 1954.

Maher, J. R., *New Perspectives in Job Enrichment*, Van Nostrand Reinhold, 1971.

Palmer, D., 'Saab axes the assembly line', *Financial Times* (23 May 1972).

Paul, W. J., and Robertson, K. B., *Learning from Job Enrichment*, ICI Limited, Central Personnel Department, 1960.

Pauling, T. P., 'Job enlargement—an experience at Philips Telecommunication of Australia Limited', *Personnel Practice Bulletin*, **24**, 3 (3 September 1968), pp. 194–196.

Work Structuring; A Summary of Experiments at N.V. Philips, Eindhoven, 1963–68, Philips Report, 1969.

Powers, J. E., 'Job enrichment; how one company overcame the obstacles', *Personnel* (May/June 1962).

Rosen, H., 'Job enlargement and its implications', *Industrial Medicine and Surgery*, **32**, 6 (1963), p. 217.

Sirota, D., 'Production and service personnel and job enrichment', *Work Study* (January 1973), pp. 9–15.

Sirota, D., and Wolfson, A. D., 'Job enrichment: what are the obstacles?', *Personnel* (May/June 1972); 'Job enrichment: surmounting the obstacles', *Personnel* (July/August 1972).

Skinner, W., 'The anachronistic factory', *Harvard Business Review* (January/February 1971), pp. 61–68.

Sorcher, M., and Meyer, H. H., 'Motivating factory employees', *Personnel*, **45** (January/February, 1968).

Tangerman, E. J., 'Every man his own inspector, every foreman his own boss at Graflex', *American Machinist*, **7**, 3 (1953).

Taylor, L. K., *Not for Bread Alone: An Appreciation of Job Enrichment*, Business Books Limited, 1972

Tuggle, G., *Job Enlargement: An Assault on Assembly Line Inefficiencies*, Industrial Engineering, February 1969.

Viteles, M. S., *Motivation and Morale in Industry*, Norton, New York, 1953.

Walker, C. R., 'The problem of the repetitive job', *Harvard Business Review*, **28**, 3 (1950), pp. 54–58.

Wharton, D., 'Removing monotony from factory jobs', *American Mercury* (October 1954), pp. 91–95.

Whyte, W. F., *Money and Motivation*, New York, Harper and Brothers, 1955.

Wilkinson, A., *A Survey of Some Western European Experiments in Motivation*, The Institute of Work Study Practitioners, 1970.

Appendix B

Brief Case Descriptions

Case	Subject	Country
A	Volvo (1) Kalmar plant	Sweden
	(2) Truck and bus manufacturer	
	(3) Proposed truck manufacturer	
B	Saab engine assembly	Sweden
C	Car-gearbox assembly	U.K.
D	Engine assembly	U.K.
E	Vehicle parts manufacture	Norway
	(1) Suspension manufacture	
	(2) Brakes	
F	Valve assembly	U.K.
G	Floor-sweeper assembly	U.K.
H	Washing-machine assembly	U.K.
J	Iron assembly	U.K.
K	Television assembly	U.K.
L	Television assembly	Holland
M	Typewriter assembly	Holland
	(1) Typewriter	
	(2) Composer	
N	Television controls assembly	Holland

CASE A MOTOR-VEHICLE PRODUCTION (VOLVO—SWEDEN)

In recent years the Volvo company has made considerable efforts to develop the manufacturing methods and systems used in virtually all of its plants. This development has been aimed at improving the working environment and increasing worker participation, and was prompted partly by the company's desire to expand and the need to recruit and retain labour in a difficult labour workout.

(1) The Kalmar plant

The layout of the plant resulted in a unique and multi-sided building, which helped solve the problem of how to build small workshops into the atmosphere of a large plant.

180

The work of assembling the components into finished cars has been split up and given to a number of teams, each of which has responsibility for its special section of the car, e.g. the electrical system, steering and controls, instrumentations, brakes and wheels, etc.

Within a team, which consists of fifteen to twenty-five persons, the members themselves can agree on how the work should be distributed and they themselves decide on when and how distribution of the work should be carried out.

Layout (Exhibit A.1—by permission of Volvo Co.)

The stores are located in the centre of the factory. The various working teams are grouped along the outside walls.

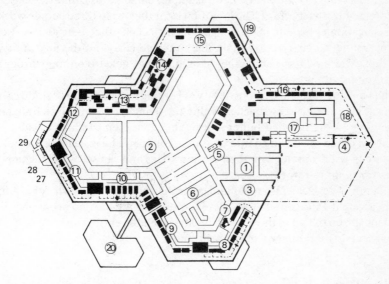

Exhibit A.1 Ground-floor plan

Painted bodies arrive by rail (4) and are taken through a washing plant (17) into the factory, where they are placed on self-propelled dollies (5). Then they pass on to a lift for transport to the upper floor, where the bodies are given interior fittings. The upper floor has principally the same layout as the ground floor.

While the bodies are being fitted out on the upper floor, the engines and gearboxes are fitted together and the front- and rear-axle assembly is built up on special dollies on the ground floor.

The bodies are taken back to the ground floor by using a lift (7) and lowered on to the chassis details.

Then assembly work continues (8–16) on the low type of dolly. Many tests and controls are carried out before the finished cars leave the factory (18).

The work teams

Each working team has its own entrance (27), changing room (29), wash room (28), break room, etc. The stores with assembly parts are located in the middle of the building. The team itself is responsible for the transport of material within the area. At the assembly bays there is storage and more materials can be called for from the stores.

In each team-sector there is generally a section for pre-assembly.

The movement of car bodies between and within the work teams is carried out on self-propelled dollies. Work on the body can either be carried out on the move or at standstill. With the assistance of a special fixture, the body, when needed, can be tilted so that work on the underbody can be carried out as conveniently as possible.

The team at 12 (Exhibit A.1), for example, has stations connected in series, surrounded by two buffers. The intention is for the two to three people working at each station to be able to exchange jobs with each other. The idea is also for the personnel to be able to follow the body and carry out the jobs at several stations. After further training they should then be able to continue through all the stations within the team.

The working team has the possibility of varying its rate of work within certain limits. Agreement can be reached within the team about working a little faster and filling up the finished buffer stock. The team then can take a pause. The members of the team can go into the break room and have a cup of coffee. But they are in good contact with their working bays and can see when the time has come for them to return to work.

Assembly work can also be organized in another way as shown, for example, by the team at 10 (Exhibit A.1). Here the stations are located in parallel. Each working group of two to three persons carries out the entire work on each body, which remains stationary while this is done.

Capacity

Total floor area of the plant (both floors) is 44,000 metres^2 and site area is 250,000 metres.^{2} The capacity of the plant is 30,000 vehicles/year from a labour force of 650, of whom 570, approximately, will be production workers working one shift. Female workers will represent approximately 30 per cent of the labour force in the new plant as in similar existing plants. The plant produces vehicles for the home market only and any expansion of the plant will be achieved by building further 'honeycomb' units from the sides of the original plant. As the original building is situated on the highest ground on the site, it will be possible to extend the building without losing window space in existing buildings.

(2) Truck and bus assembly

1500 people are employed at the main truck-assembly plant, of whom 750 are direct production workers, including 5–10 per cent females and 20 per cent immigrants. Five basic bus and twenty basic truck types are produced at a rate of approximately 15,000 units per year.

Attempts to introduce 'team working' in the plant began around 1971 with the introduction of experimental systems in certain pre-assembly areas, i.e. areas off one of the four main vehicle-assembly lines in the plant, each of which are manned by up to 100 workers.

The principal objective of these 'experiments' was to delegate increased responsibility to workers, and in order to do this teams were established to be together responsible for the work undertaken in their area. Each team selected a spokesman who may change periodically, and in general teams consisted of between six and twelve workers, although some were as small as three workers.

Wheel/tyre pre-assembly

A typical example of team working in a pre-assembly area is provided by wheel and tyre assembly.

The wheel/tyre pre-assembly consists of ten workers, one of whom is selected by the team as a spokesman for communications with foreman, material handling, etc. In general the spokesman also plans the work for the group, having been given a weekly production schedule. Wheels are delivered to each of the four assembly lines (at the requisite rate) and to packing and spares departments. The production schedule shows the hourly output required for all types of wheel, requirements for the lines being closely specified whilst requirements for the other two areas are more flexible in that they may be satisfied as convenient to the group, rather than at a given regular rate. Three of the workers in this group work on a flow line, and it is obligatory that operatives on these jobs move to non-line jobs at regular intervals. Otherwise workers are free to rotate jobs if they choose to do so.

Replacements for team workers are obtained by the foreman who requests labour through an internal agency. In the event of absenteeism, the spokesman, representing the group, either requests extra labour through the foreman, or indicates that the team will cover for the absentee, occasionally with appropriate schedule changes. The plant carries 12·5 per cent extra labour to cover against absenteeism.

The cycle time for workers in pre-assembly teams is generally near 10–15 minutes, compared to 25 minutes on the line, which in turn compares with the 2–3 minutes normal on car-assembly lines in the company.

Monthly meetings are held between spokesman, foreman and the engineer responsible for the area, to discuss problems. Independent quality inspectors examine all output, often when pre-assembled items are incorporated in vehicles on final assembly.

Final assembly

Team work has also been introduced on parts of the vehicle final-assembly lines in the plant. The teams consist of four to six workers at adjacent sections of the moving-belt, fixed-item line. The cycle time of the line is 25 minutes and

184

workers autonomy covers allocation of work, job rotation and quality responsibility. The workers in the teams may decide to work together or to work on separate tasks, and are free to rotate as required. The lines contain one item, i.e. vehicle chassis, per team rather than one per operator, thus the teams consist of all operators multi-manning each station. There are eight such teams on the line, stations (i.e. teams) are not separated by buffer stocks, no team representative is elected (or required?) and teams do not have regular meetings with supervision. Responsibility for quality derives simply from the fact that faults identified on inspection are traceable to work teams. Workers in all such groups are paid on a group piecework system.

(3) Proposed truck-assembly method

In 1971 a study of truck-assembly methods was undertaken. An average cycle time of 25 minutes (as at present) was established, and the configuration shown in Exhibit A.2 was proposed.

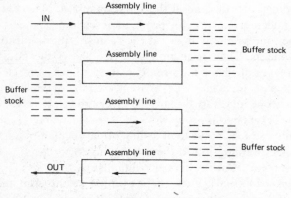

Exhibit A.2 Truck-assembly line

Several models of truck were to be assembled. Each small assembly line was to have three stations. Each station was to be manned by approximately six workers, one truck being available at each station. It was not intended to employ moving-belt-type lines but rather indexing lines with the indexing movement controlled by signals from the stations on the line. Buffer stocks were to be provided between lines, the cost, i.e. investment, of such stocks being considered to be low compared with the stocks of finished trucks at the end of production. Pre-assembly work was to be undertaken adjacent to the area of the flow lines.

CASE B   SAAB CAR-ENGINE ASSEMBLY (SWEDEN)

A unique approach to the assembly of a range of petrol engines was adopted by the company following extensive investigation, prior experimentation and consultation. The following account provides a very brief description of the new approach and the background to its introduction. (Exhibits by permission Sabb Co.)

Design of method

A fundamental premise was that 110,000 engines were to be produced annually. It was also known that the engine was to have relatively few components, that they were light and that the engines were to be manufactured in only a few versions. Women were to be employed in addition to males, indeed approximately 50 per cent of the work force was female. A large amount of production engineering material was also available from the original manufacturers of the engines. In fact, it would have basically been possible to transfer the production methods used originally without a great deal of extra work. However, it was considered to be desirable to seek a method that would reduce the extent to which an individual was tied to his job and increase his possibilities of developing with his work tasks. It was considered to be desirable that assembly workers should be able to influence the distribution of work and their own work considitions.'

'Nevertheless, the use of a conventional assembly line was a real possibility throughout the entire project stage. In other words, it was not a foregone conclusion that there would be a solution other than this. All possible methods were analysed. The combination eventually chosen was: *group assembly, undriven system at the actual assembly points, engine moved to material, fitters accompany the engine, material supplied in bulk* (entire pallets are transferred by truck and located at the assembly points in racks, each pallet containing only one type of item). In addition to *handling by truck, conveyor handling* is employed for transporting engines to and from the assembly groups.'

'The next development was to visualize a number of similar groups working in parallel (Exhibit B.2). This alternative was employed in subsequent development work. The next stage was to organize transport to and from the groups in a conventional manner. The pre-assembled components to be supplied and the completed engines which were to be transported out would both be carried on a common transport system. A large conveyor loop was designed, between the long sides of which the assembly groups were located

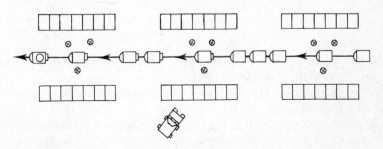

Exhibit B.1

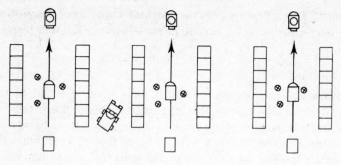

Exhibit B.2

transversely. This achieved a system which met the stringent requirements of a work-saving transport system, independent work tasks, minimum ties to the work and maximum scope of assembly.'

'Only one problem now remained. The truck drivers would be compelled to cross the mechanized conveyor loop in bringing materials to the racks. It would be far too complicated to install a conveyor which would cross the transport ways at sufficient height to allow the trucks in and, even though this might be technically feasible, it would be too expensive. The ultimate solution was to locate the oblong buffer loop completely to the side of the groups and to furnish each group with a U-shaped guide track in the floor. In this way, the trucks were allowed free passage, while the other advantages were maintained' (Exhibit B.3).

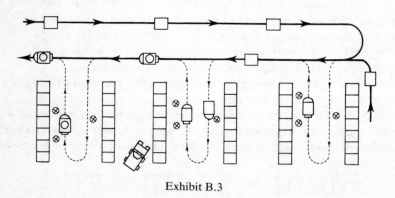

Exhibit B.3

The new engine factory was built to these blueprints (Exhibit B.4) and was commissioned in the Autumn of 1972.

The layout and method enables each assembly group to complete an engine from start to finish, apart from pre-assembly components. Of the 7 assembly groups one is intended as a training group. Each group may determine how the work is to be distributed within the group, and there is no mechanical control during the actual assembly work. The maximum cycle time for a worker—when

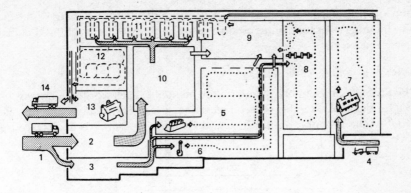

1   Goods reception

2   Arrival inspection purchased factory parts

3   Raw material store

4   Engine blocks (material) from own foundry

5   Machining cylinder heads

6   Machining connecting rods

7   Machining engine blocks

8   Machining crankshafts

9   Pre-assembly

10   Parts store

11   Group assembly

12   Engine testing

13   Ready stock

14   Engines to Trollhättan and Nystad

Exhibit B.4

assembling a whole engine—is 30 minutes instead of the 1·8 minutes as would have been the case with an assembly line. The extent to which job 'enlargement' is utilized depends mainly on the ambitions and knowhow of the fitters themselves, thus working individually a worker has 30 minutes for her job, whilst working in pairs gives a 15 minute job, etc.

CASE C   CAR-GEARBOX ASSEMBLY (U.K.)

This case describes one method of gearbox assembly in one large plant in the motor industry in the U.K. Gearboxes for a range of vehicles are manufactured in the plant. Basically such vehicles are grouped as heavy, medium and small passenger vehicles. This case relates to the assembly of gearboxes for small vehicles.

All workers are paid on a day-rate basis coupled with a job-grading scheme, with associated payment grades. Two-shift working is employed throughout.

Shop supervision is through foreman to superintendent. Training is 'on-the-job', 'floating' workers being used to assist all trainees until required performance is achieved.

Gearbox-assembly 'carousel'

Gearbox final assembly is undertaken manually, each worker being responsible for the final assembly of a *complete* box, given various subassemblies. The layout of the assembly area is shown in Exhibit C.1. The overhead 'carousel'

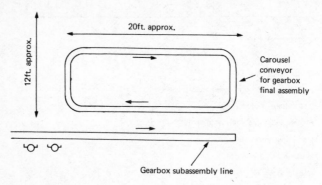

Exhibit C.1 Layout of gearbox final-assembly area

conveyor transports twelve equally spaced hanging carriers continuously in the direction indicated. Each carrier provides a single work station and consists of two trays of components, e.g. nuts, gaskets, etc., replenished from stores by a 'feeder' worker. Beneath, and in front of, these trays is located a work-holding device to which a gearbox may be clamped and rotated in a vertical plane, and locked in either horizontal or vertical position. Suspended in fixed positions, beside this continually moving overhead carousel conveyor, are five pieces of equipment required in the assembly process, manually controlled nut runners, air-powered tools, pneumatic hand-operated press and power screwdriver.

Adjacent to this moving carousel conveyor is the final portion of the flow line on which certain gearbox subassemblies are made. This line supplies, via a sequence of assembly stations, the principal subassemblies, case, etc. for the gearbox final-assembly carousel. Finished gearboxes are transported by truck and pallet away from the carousel area.

Approximately twenty assembly operations are performed by each worker in assembling a gearbox carousel. In performing these operations workers walk round the work area at a pace determined by the speed of the carousel. The fixed equipment is located where required and one complete 'circuit' of one carrier on the Carousel, and hence the maximum time available for gearbox assembly occupies 6 minutes 22 seconds. In comparison, the cycle time at stations on the flow lines employed for the assembly of gearboxes for medium cars, ranges from 35 seconds to 48 seconds, and for heavy cars the lowest cycle time is 1·20 seconds, approximately.

CASE D   CAR-ENGINE ASSEMBLY (U.K.)

The new plant referred to in this case, one of several operated by a major manufacturer, is devoted entirely to the manufacture of engine/gearbox units for a range of medium-price family cars. Three main assembly lines are used in this highly mechanized plant, and all are employed on batch work. Lines 1 and 2 have a cycle time of 3 minutes, whilst a cycle time of 4·16 minutes applies to the third line. Lines 1 and 2 were designed for a cycle time of 2·6 minutes and previously operated at 3·6 minutes. All three lines are moving-belt, automatically *indexing* lines designed to incorporate a substantial amount of automatic equipment. There are forty-one workers per line per shift working individually except in the case of piston and big-end assembly.

Engine blocks are fed on to separate pallets on the raised line, which are indexed automatically on completion of the cycle time (there is no control of the indexing from stations) by mechanical means. Equipment, parts, subassemblies, etc., are fed to stations by marshallers, and because of the nature of the line operation workers are closely paced, confined to their stations and entirely dependent not only upon prior operations but also upon the indexing mechanization of the line.

A substantial amount of subassembly work is undertaken off the line, occasionally on feeder assembly lines. For example, crank balancing and clutch assembly forms a three-station feeder line parallel to the main line. Connecting rod, piston and big ends are supplied complete to the line, hence big-ends must be removed prior to assembly on the line.

Several fully automatic stations were employed on the lines, e.g.

> Tightening of main-bearing nuts, i.e. after nuts had been 'started' on studs at previous station and prior to manual tightening to required torque at next station.
> Tightening of head nuts (as above).
> Tightening of flywheel-housing nut.
> Tightening of clutch cover.

In each of the above cases the assembly machine occupies a whole work station (about five feet of the line) and operates on a work-time of considerably less than the line cycle time; often only a few seconds of work are performed and hence the equipment and station are idle for a large proportion of the time

The mechanized equipment employed on the lines had not operated entirely satisfactorily, nor were the company entirely satisfied with the indexing lines. In general, it was felt that it might be desirable to abandon the automatic stations on these lines in order to

(1) Provide more space on the lines, e.g. by using manually operated overhead-slung equipment.
(2) Provide less dependence on machinery which had proved to be unreliable.

CASE E   MANUFACTURE OF VEHICLE PARTS (NORWAY)

The following account outlines three of the mass-production methods employed in one plant of a large and diversified engineering company in Norway. The cases relate to motor-vehicle-component manufacture—a minor, atypical, but growing aspect of this company's business.

(1) Suspension link-arm manufacture

The item sketched in Exhibit E.1 forms part of the steering/suspension of certain Swedish motor cars. The item is forged and pressed in one department of the plant prior to being fully machined elsewhere.

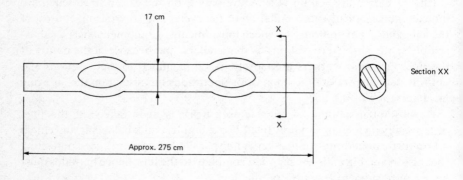

Exhibit E.1 Suspension link arm

These arms are forged and pressed at a rate of 300 per hour by a team of three semi-skilled male operators. Three items of equipment are employed. These are, in manufacturing order—an induction furnace in which bars are heated prior to forging on a power hammer, and finally, a press having two 'stations', the first for the removal of 'flash' from the forged part and the second for pressing the two parallel sets or 'flats' on the item. These three machines are linked by roller conveyor, and the operators at each machine are required to handle items with tongs and operate foot-pedals. The items have been manufactured in this manner for almost seven years. Operators have, of their own volition, adopted a system of job rotation in which each operator changes tasks within the group approximately every twenty minutes. However, it has been noticed that despite this virtually regular job rotation each operator is considered to be a specialist on one of the three machines, hence should any adjustment or any action out of the ordinary be required, specialist operators tend to return to their machines.

One reason for the adoption of a system of job rotation is the unbalance of the labour content of the three operations. In particular, the furnace operation requires relatively little manual work, hence the operator on that machine, whilst working at the same cycle time as the other two operators, spends far more of his time idle. For this reason the two other operators are not provided with a relaxation allowance in their work times. (A relaxation allowance of 6 per cent is normal on other operations.) The furnace operator is expected to help out his colleagues in respect of contingencies and matters for which time allowance is normally provided. However, the probability of job rotation was also borne in mind when designing the tasks.

The workers are paid as a group on a piecework incentive. All output is paid for, there being no inspection of products after this stage of the manufacturing process.

(2) Machining of brackets for brakes

Present method

A complex cast bracket for truck brakes is presently completely machined by three semi-skilled workers, working one shift. Cycle time is approximately 2·75 minutes. The facilities employed in the machining of the item are shown in Exhibit E.2. Provision is made for buffer stocks between machines. One

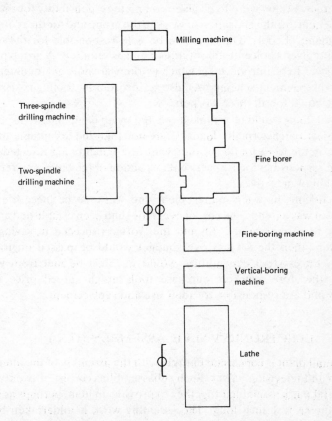

Exhibit E.2 Equipment for machining of bracket

worker operates the lathe whilst the other two workers, who change and share jobs, perform all remaining operations on the other machines.

Workers are paid on an individual piecework system, and whilst an independent inspector examines a sample of completed items, workers are paid for all output.

Proposed method

It was (1973) proposed to revise the method of working employed for the manufacture of this item. No changes would be made in the layout or nature of the equipment; however, the degree of autonomy of the group of three workers was to be increased.

The motives for this change, as envisaged by management, were

(1) To increase the degree of 'interest' in the job.
(2) To reduce muscular fatigue of workers.
(3) To enable workers to increase their earnings.
(4) To try to save expenditure on tools.
(5) To save money in terms of unit costs, by paying for only 'good' output.

The three workers were to be given complete responsibility for the machining of the item, for the allocation of work in the group and for the requisitioning and changing of tools. They were to be held responsible for the quality of output and given choice of job rotation if and as required. A group piecework incentive was to be introduced in which workers are not paid for detected 'bad' output. The payment scheme was also to incorporate tooling costs. Payment was to be equal for all three workers.

Finished items would be sample inspected by an independent inspector who would reject batches (pallet loads) if greater than the acceptable number of defective items were found. In the event of a batch being rejected, workers would be responsible for 100 per cent inspection of the batch and rectification or rejection of defectives.

In discussions between management and workers the piece life of every cutting tool was agreed. The cost of every tool had been established and a piece price for tools determined. Should the workers succeed in saving on tool expenditure then the savings over budget would be passed on to them as earnings. Excess tool expenditure would in effect be met from wages. In practice, therefore, workers purchase tools at an agreed price from the company and are responsible for their use and replacement.

CASE F   ELECTRONIC-VALVE ASSEMBLY (U.K.)

This small plant is concerned entirely with the assembly of the internal parts for radio and television valves. Such subassemblies consist of twenty or thirty small metal parts assembled together to provide an item, perhaps as small as $\frac{1}{2}$ inch diameter $\times$ 1 inch long. The assembly work is undertaken by female workers, working with jigs, simple hand tools, tweezers, simple presses, etc.

Such work is undertaken at work benches under closely controlled conditions, cleanliness being particularly important. A variety of types of valve are made, valves in general being classified according to the number of stages involved in their assembly, i.e. two stages (two assembly operations) to seven stages.

Traditional production method

Valves are traditionally assembled using a system which approximates to a non-mechanical flow line. Women workers, each undertaking one of the operations in the assembly of valves, are divided by substantial buffer stocks. As the time required for each operation varies, some paralleling of stations is required in order to achieve balance between operations. Workers, seated at work benches, assemble several valves at once, each located in a single jig at the work-place. Boxes of semi-finished valves are then transported to the next operation, etc.

Revised production methods

The need for a possible change in production methods was first suggested in connection with a high absenteeism and labour turnover evident in the plant. Following an attitude survey, a series of meetings between management and workers and the examination of alternative production methods, it was decided to experiment with new approaches to valve assembly. A group of eleven self-elected operatives, following discussions, began to work as an independent group shortly afterwards. Following the decision to establish this volunteer group a period of training was provided to ensure labour flexibility, and prior to this the group, as their first group decision, were required to determine the layout of the benches in their new work area, which was separate from the existing assembly areas.

The group were established as a semi-autonomous work group, their autonomy and responsibility being extended in comparison to that which applied in the traditional methods of valve assembly to encompass the following:

(1) Output targets were fixed using the same work standards as had applied in the previous system of work, i.e. the same output was required from workers. However, the workers, who were paid on a form of individual incentive system, were able as a group to determine how the output requirements were to be met, insomuch as when an individual left the group they were able to decide whether or not to ask for a replacement for the worker, or whether to increase their own performance levels to make up the necessary output.

(2) Members of the group were each responsible for accepting or rejecting components supplied to them. It was usual for the group to nominate a representative who would liaise with supplier departments on issues relating to quality of parts supplied to the group.

(3) New members for the group were initially trained for a period of one month in the separate area of the plant. This initial basic training was followed by a period of training within the group. The group were responsible for arranging this period of further training, the group being required to call in trainers should any problem occur. The group were also able to influence the training given to new recruits whilst undergoing their one-month basic-training period, e.g. dependent on their requirements, and their own performances, they were able to specify the type of training provided to the individual, e.g. the operation on which that individual would initially be trained. The period of further training within the group was intended as speed training, and training on a different operation in order to maintain group flexibility.

(4) Groups were responsible for their own quality control. Items assembled at stations were checked by the assemblers. Final assemblies were either dispatched directly to the customer plant or if any doubt existed, for example when faulty material had been received, the group would nominate a quality controller whose task was to sample the final product. This task would normally be rotated amongst the members of the group. The need for rectification and repair would also be decided in the group, similarly the group was also responsible for the decision as to whether or not items should be rectified or scrapped. If the scrap level from the group was considered to be too high the group were able to request a meeting with management in order to identify the reasons for the high defective rate.

(5) The group was required to handle their own weekly performance-assessment meetings. They were required to nominate an individual to meet regularly with management to discuss output, and decide on any necessary corrective action. Certain paperwork was also to be prepared for the purposes of performance assessment, the group being required to complete documents showing total output, rejects, rectifications for each individual.

(6) In addition to the above paperwork, the group was also required to make out requests for maintenance and repair work, advance orders from stores for parts, etc.

CASE G   FLOOR-SWEEPER ASSEMBLY (U.K.)

The company manufactures domestic houseware appliances, including kitchen equipment and other appliances, largely of a non-electrical nature. The plant referred to in this case employs 600 persons of whom 400 are hourly-paid production workers. Floor sweepers are the principal output of the plant. The whole range of sweepers are of the non-powered, 'push-and-pull' variety and consist essentially of a long handle; case (metal or wood) wheels and associated rotating brush(es); dust pan(s); together with chassis frame, suspension units, adjustment mechanism, trim strip, etc. Currently ten models are manufactured.

Previous production methods

Prior to 1968 wooden sweepers were assembled on twenty-seven, female-operative, moving-belt, removable-item flow lines. After this date such lines were replaced by similar lines manned by ten operatives. Similarly, the assembly of metal sweepers was modified, being achieved by fourteen-operator lines prior to 1969, and eight-operator lines from that date until the most recent changes. In both cases, the changes were effected primarily in order to improve labour utilization through a reduction in line-balancing loss. A typical eight-operator line operates with a cycle time of 35–45 seconds, and a balancing loss of 9 per cent.

New production method

Over a ten-month period the present method of assembling metal sweepers was introduced. The new method relied upon teams of two assembly operatives; provision was made for fifteen pairs of operatives to replace the previous assembly lines for these models. Each operative worked in a standing position at a bench, equipped with the appropriate jigs, tools, and supply of parts. The operative on the first operation passed semi-finished items to the second operative, from where they were transported by belt conveyor as previously. In general the average balancing loss of such work teams is 2 per cent and cycle time 2·4 minutes, i.e. an output of 25 per hour. Women workers are employed, in many cases the same workers as previously engaged on the flow lines. Payment is on a piecework basis, as previously.

This method of working was introduced as a means of increasing labour utilization through reduced balancing loss. The introduction of some degree of job enlargement, and possible enrichment, was seen as a possible consequence of the introduction of team working but was not seen as a major benefit, nor used in the assessment or advocacy of the new method. The move to team working, with increased cycle time, was seen as a logical extension to previous changes, also aimed at increased efficiency through reduced balancing loss.

Individual build, i.e. one operative/one sweeper, was rejected as unpractical, largely because of the size of the items being assembled, the size of the jigs and rack space required, and the confined space available for work benches. Additionally, whilst the present method of working necessitates the provision of some process allowance to cover balance delay for one operator, considerably more allowance was thought to be necessary to cover operator movement at the work-place had individual build been adopted.

CASE H WASHING-MACHINE MANUFACTURE (U.K.)

The description below relates to the manufacture of drums for domestic automatic washing machines in a plant employing 700 workers.

Assembly of drum for automatic washing machine

The metal drum for the automatic washing machine is assembled by a six-man team. The drum is assembled from various small metal pressings, e.g. the hinged doors, and several small plastic mouldings, all of which are attached to the body of the drum which is pressed from strip metal. The washing machine drum is approximately thirty inches in diameter and fifteen inches long.

The equipment employed is shown in Exhibit H.1, whilst operations are as follows (figures refer to Exhibit H.1):

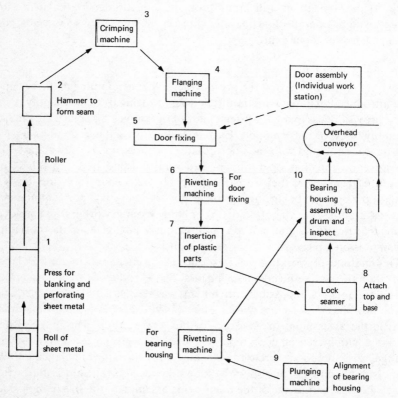

Exhibit H.1 Arrangement of facilities/operations in drum assembly

(1) The strip metal is blanked and perforated on the mechanical press. Each blank is then rolled to form the cylindrical shape.
(2) The two ends of the roller metal are brought together and hammered to form a seam.
(3) The seam is crimped.
(4) Flanges are formed at the top and bottom of the tube.
(5) The doors are attached over the opening in the cylinder wall.

(6) These doors are rivetted to the wrapper.

(7) Several plastic members are inserted into the metal tube.

(8) The top and bottom of the drum (pressed elsewhere) are lock-seamed to the metal tube.

(9) The drive-bearing housing assembly is made up.

(10) The drive-bearing housing is married to the drum and the completed assembly inspected before being attached to the overhead conveyor for transportation to the automatic-washer assembly line.

Payment is by group piecework. A job-grading scheme is employed to determine basic pay grade. Ability to perform more than two jobs entitles the operatives to the higher pay associated with a higher job grading.

The group of workers organize the work between themselves. Work rotation is informally organized within the group. The longest-serving group member is the leading hand who is expected to take to the chargehand any problem unsolvable within the group.

The group are responsible for the quality of the end product. Any assembly which is found to be substandard on inspection, or at a later stage, is returned to the work group to be made good. Usually three of the group spend approximately one hour on Friday afternoons on such rectification work. The men operating the rivetting machines are responsible for some periodic resetting work. Machine breakdowns or other peculiarities are to be reported to the chargehand. The men, therefore, organize the work so as to minimize the disruptive effect of these breakdowns by building up stocks of up to 500 completed blanks to be used whilst repairs are carried out.

Any increase or change in manning levels or output requirements are undertaken in consultation with the work group. On the last occasion that the group was enlarged difficulties in achieving previous output levels were experienced. It was noted that the press was the limiting stage. The group discussed this with the supervisor and decided amongst themselves who was the most efficient press operator and persuaded him to do this task.

The group have regular bi-monthly meetings with the section supervisor in his office to discuss their problems and talk about possible improvements. The supervisor also stressed the importance of his regular contact with the group at the work-place. The members of the team are carefully selected. The policy adopted is that of transferring, from within the department, workers known to be suitable for group work rather than introducing new starters directly into the group. The tasks performed by members of the group are not difficult to learn but training time to reach the skill level required varies from six to eight weeks.

CASE J   ASSEMBLY OF STEAM IRONS (U.K.)

This case relates to the final assembly of electric steam irons for domestic use. A range of approximately twenty irons are manufactured, all basically of the same design but having various colour, voltage and style specifications. Each iron consists of six main subassemblies, i.e. sole plate with switch and thermostat, cord, handle, shell, rear door and pump.

Previous production method

The final assembly of the irons was previously undertaken on two manual non-mechanical-type flow lines. The principal line was staffed by eight unskilled female workers—each of whom assembled part of the product, working on four items held in a tray on the work bench. These trays of four irons ran on tracks along the line, and there was no provision (no space) for buffer stocks of trays of partially completed irons between stations. Thus the stations on the line, whilst not mechanically paced, were required to index virtually simultaneously. Workers on the line were paid on a base rate and group-output bonus. All parts were located in front of and around the work stations, items being supplied to the line by feeders. A further four-station line was also operated on exactly the same principle.

New method

The cycle time of the eight-station line at standard performance had been fixed at 43 seconds after time study. However, workers on the eight-station line consistently gave a very high performance against this standard.

An MTM 2 analysis of the final assembly task gave a total work time of 3·49 SM compared to the original time of 5·74 SM. Such an order of reduction was not considered to be possible with the existing assembly system, hence, in order to establish new work standards and to reduce interoperation handling time, a system of individual assembly was introduced. Assembly stations were arranged adjacent to a two-level gravity-fed roller conveyor. Eight individual assembly stations were used.

A standard time of 3·49 minutes was given and initial output from eight benches was approximately 800 per day. Following discussion with workers the work standard increased to 3·67 SM and over a period of four months output from the line increased to a figure equivalent to a performance of a little over 100 BSI.

Workers are supplied with trays containing the necessary subassemblies, all small items (screws, springs, etc.) being supplied to the work stations. Irons subsequently found to be defective for reasons attributable to faulty assembly are returned to the appropriate worker, who is responsible for rectification 'in her own time'—i.e. no additional time allowance is given. Workers operate on an individual piecework system and irons are assembled in batches.

CASE K ASSEMBLY OF TELEVISIONS (U.K.)

Previous assembly method

The final assembly of monochrome television sets, i.e. assembly of tube to cabinet, components to cabinet, test, pack, etc., was previously undertaken on approximately five moving-belt assembly lines. Each line operated at a cycle time of 1·75 minutes and was staffed by approximately thirty employees, twenty of whom were female assemblers, the remaining ten being males engaged on testing, packing, loading, etc. The moving-belt line was fed from an overhead conveyor from which most major components were supplied to the first station on the moving belt. The remaining assembly stations were each supplied with components, subassemblies, etc. and the line operated with approximately three times as many sets on the conveyor as stations, hence the assemblers, who worked in a standing position, were able to move out of their station without this resulting in immediate 'blocking' or 'starving' of adjacent stations. The lines were balanced manually, balancing loss was thought to average 10–15 per cent. Workers on the lines were paid on a piecework payment system geared to the line, or group, output. Televisions were assembled in batches on these lines, up to twenty different types being manufactured.

New assembly method

Several of these belt lines were arranged in the configuration of an 'E'. Initially each line was supplied with tubes and cabinets brought on pallets by handtrucks from stores. Each line employed seven operators, including quality and rectification workers. The products were passed between operators by sliding along the bench, no form of mechanical pacing or handling being employed. There was space for a maximum of two items in buffer stocks between stations, and three such configurations were used to replace one previous moving-belt assembly line. The new lines were used primarily for the assembly of items manufactured in small batches, e.g. export models, and were designed for a cycle time of approximately 5 minutes.

Following the introduction of these lines some difficulty was experienced in achieving the desired output, hence some modifications were introduced primarily in relation to the materials supply to the lines. In the present arrangement, two 'E' type lines serve one packing line, and conveyor linkage had been introduced between the end of the assembly line and the packing line, otherwise the method of working in the configuration of the assembly area was identical to that discussed above.

The new lines were introduced primarily to increase flexibility, and to a lesser degree to increase quality during assembly. The principal advantages seen of the line were the increased flexibility, hence the lines were eventually employed for the manufacture of two-thirds of the model range. A few disadvantages were claimed for the lines, it being conceded that some job enlargement had taken place, although little enrichment was introduced. Behavioural advantages were not an important factor in the introduction of the new lines, although in retrospect it was considered that the smaller working groups, in comparison to the previous thirty-man lines, offered some of the advantages often associated with small group working.

CASE L TELEVISION ASSEMBLY (HOLLAND)

Complete assembly of a monochrome television set in this plant consists basically of the following operations:

(1) Pre-assembly and pre-preparation, e.g. wiring harnesses.
(2) Printed-circuit-board assembly
(3) Soldering of circuit boards.
(4) Touch-up and final assembly of components in circuit boards.
(5) Fault find, test and adjust circuitry on boards.
(6) Assembly of tube speaker controls, subassemblies and circuit boards to cabinet.
(7) Test and adjust.
(8) Soak, i.e. operation of unit for several hours.
(9) Inspect.
(10) Finish and pack.

These tasks were previously carried out by both individual and flow-line assembly methods. Pre-assembly (1) was undertaken by females working individually, whilst circuit-board work (2) was undertaken on a flow line. Wire crop, automatic flux and solder (3) was undertaken by one operator. The assembly of further, normally larger, components (4) was then undertaken before the board was passed for test and adjustment. Assembly of the items into the cabinet was undertaken by male workers on a non-mechanical flow line whilst further male workers were responsible for test and adjust, soak, test and packing. In total, thirty workers were associated with the television assembly-line working at a cycle time of 4 minutes.

New assembly method

As an experiment, a group of seven workers was established as an alternative to the above method. These seven workers together undertook all of the operations accomplished by the thirty operators and worked at a cycle time of 25 minutes. The basic operations taken in the group are as follows:

(1) (a) Pre-assembly.
    (b) Printed-circuit-board assembly (50 minutes).
    (c) Solder.

(2) Touch up.
    (b) Final board assembly.

(3) Fault find.
(4) Assembly of television.
(5) Test and adjust, sock and pack.

Two female workers were in general to be employed on operation (1), otherwise each operation was to be performed by one operator, with operations (3), (4) and (5) manned by males. A seventh worker (male) covered for absenteeism in the group, or helped out in the case of no absenteeism. The group were located together in an area of the plant (Exhibit L.1) and were allowed to devise their own work procedures, job rotation, etc. The division of the job into tasks was specified for the group although workers were to

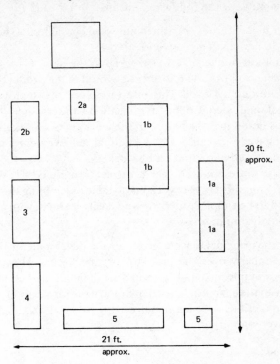

Exhibit L.1 Layout for television assembly group

determine whether they completed one item at each operation, or whether they worked on batches of items.

Workers were paid on a day-rate system, pay being fixed according to a grading scheme. No grade increase was given on introduction of the system. The group were to undertake their own quality control, a target output quality being established in consultation with management. Output was sample-inspected on a batch basis.

Following establishment of the group, supervision was provided only during group consultation on problems that arose and on the setting of standards. The group were given direct functional contact with the staff of service and technical departments and they were responsible for calling up materials by telephone. Workers were responsible for the completion of paperwork.

Workers in the group were asked to establish their own output standards and they were to report straight to their departmental head. Members of both groups were chosen following discussions with workers on existing lines and close supervision was provided during the setting-up periods.

CASE M   ASSEMBLY OF ELECTRIC TYPEWRITERS (HOLLAND)

Typewriter assembly

The company had chosen to examine their existing methods for electric-typewriter assembly for the following reasons:

(1) Absenteeism and turnover in the plant were considered to be high (12 per cent and 30 per cent respectively).
(2) A large amount of overtime was worked in the plant.
(3) Considerable expense and effort was devoted to the production of items at the requisite quality level. This often involved the deployment of large numbers of quality controllers, rectification workers, etc.
(4) It was considered that the nature of the work force was likely to change in the coming years because of educational achievements, and pressures towards increased participation by workers.
(5) Labour cost represented 90 per cent of assembly cost (excluding overheads, etc.) but most of the work involved in assembling the products was considered to be simple, repetitive and easily learned, with a cycle time of approximately 3 minutes.

These were the principal reasons given for the developments that had taken place in the company over the past two or three years. However it is also emphasized that simple, economic production demands had also given considerable impetus to the changes. In particular the need for flexibility in a situation where output quantities were continually being changed, and improved quality performance were important.

Previous methods

Electric typewriters had previously been manufactured on two parallel non-mechanical flow lines, staffed by both male and female workers. Approximately sixty-five people were engaged on each line, each line being divided into five main sections, the sections at the beginning of the line being devoted to more complex tasks, whilst those towards the end of the line were concerned with more simple assembly tasks. The section at the end of the line was concerned with most of the quality control and rectification work. These sections were divided by large buffer stocks, and furthermore the stations within each of these sections were also divided by buffer stocks of two to three items. The workers in each of these sections were supervised by one foreman, although each foreman was not located adjacent to the line but rather in a separate area.

Some difficulty was experienced with these lines. Firstly, problems were experienced when output was to be increased, and output had been increased at regular intervals over the past two or three years. This involved manual rebalancing of the lines. Some retraining was also normally necessary and output following such changes was not entirely predictable. These lines were engaged on the assembly of two basic models, although over 400 varieties were available, differences resulting from changes in colour, typeface, electrical specification, etc. Models were therefore made in extremely small batches. All items were made to order rather than to stock, hence a batching policy was not available.

Quality problems were experienced, perhaps because the assemblers on the lines tended to rely on the quality control and rectification workers at the end of the line to build in the product quality. It was not possible to identify faults with assemblers since faults could be caused at subsequent stations during the assembly of further parts of the machine. Communication between the various parts of the line and between these lines and the supervisors and the technical staff was said to be poor.

The payment system employed at the time consisted of a base rate fixed by a job-evaluation system, plus an individual bonus payment which might account for up to 25 per cent of the total pay.

New method

Evaluation of the situation on these two non-mechanical lines, together with the examination of the results of the attitude surveys regularly conducted in the plant, led management to believe that smaller work groups were required, preferably small independent departments. Such an arrangement was considered to offer the possibility of closer cooperation, greater worker indentification with the product, and therefore better quality, easier control and greater flexibility.

The first 'mini-assembly' line, in which twenty assembly workers were occupied, was introduced in 1971. Final assembly was undertaken on the line, subassembly being provided elsewhere in the plant (e.g. keyboard, bar-frame, motor group, etc.). The cycle time on the line was to be approximately 20 minutes (compared to 3 minutes on the previous line). Both men and women were employed (equal pay was provided for men and women in the plant), women being employed on simpler and lighter tasks whilst men were employed on more of the final tasks which involved lifting and moving heavy items. The layout of the line is shown in Exhibit M.1. Roller conveyors were used, hence the line operated in a non-mechanical fashion, there being sufficient capacity for up to two or three items between stations. All items, parts, subassemblies, etc. were stored underneath the line and various components and pieces were available on benches beside operators. The manager or foreman in charge of the group was also located in the same area.

Such mini-lines were eventually established to replace both of the previous flow lines. The system afforded increased production flexibility, hence increased output requirements could be satisfied merely by the addition of one or two more lines, without affecting any existing group. Quality control, as a separate function, was eliminated, assemblers being given responsibility for their own quality. To some degree the operations being performed on the lines had been specified by management in order that each should be a self-contained unit, and that the work of one assembler would not subsequently be

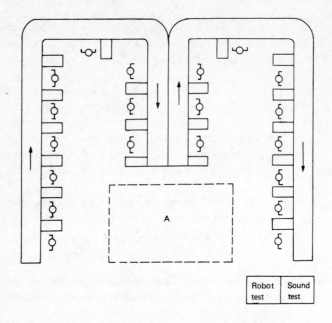

Exhibit M.1 Mini-line for typewriter assembly

affected or the product damaged by subsequent operations. Hence faults at the end of the line were readily identified, products were returned to operators. Coincidental with the introduction of the mini-lines a monthly time-rate pay was introduced for all operatives, hence workers were removed from any form of incentive pay system. Payment was based on a grading established by job evaluation, and regular performance review.

Prior to transferring operators from the existing lines onto the mini-lines, a period of additional training was provided in two-to-three-hour sessions twice per week, partly in work-hours and partly during overtime working.

Composer assembly

The composer unit, resembling a large and more complicated electric typewriter, is assembled entirely by a group of thirty male and female workers: the group operates almost entirely independently, and is a self-supporting unit.

The group is located in one corner of the assembly plant, there being basically two areas, one devoted to the final assembly of the items and the second concerned with the manufacture of the various subassemblies.

The final-assembly area operates on a non-mechanical flow line, with provision for buffer stocks between stations. Twelve operators, mainly males are employed in this area working on tasks with cycle times varying from $\frac{1}{2}$ minute to 52 minutes. Most of the operators in the area are capable of performing between two and seven operations. The subassembly area in which more female workers are employed has itself been divided informally by the workers into several smaller areas concerned with, for example, the keyboards of the machine, in which three workers are employed on a task which takes a total of 1·2 hours. In total there are eighty operations to be performed in a subassembly area, ranging in cycle time from 1 minute to 1 hour. Fifteen workers are employed in this area. Within the final-assembly area several work benches are arranged on which extremely-short-cycle repetitive tasks must be performed. These benches are not continually manned, but workers whose jobs are interrupted, or delayed, by for example a shortage of materials, are required to move to this area as necessary in order to manufacture the necessary number of parts. The manner in which workers move to the area, the number of parts manufactured, batch sizes, etc., is entirely arranged by the workers themselves, as is the deployment of the labour in the subassembly area and in the final area. Job rotation in both areas and between areas is organized by the group, and the group of both the final workers and the subassembly workers can operate against a weekly requirements list. The requirements for each subassembly and each final-assembly item are posted in the group; workers operate against this list, determining their own batch sizes, etc. Workers on these lines are responsible for their own quality, no inspection operations being included. Most of the people in the area were able to do 60–70 per cent of the jobs on the line. Subassembly workers were required to take completed items to the stores, final-assembly workers being required to draw subassemblies and items from stores.

A certain amount of vertical integration had taken place in the area in that some of the assembly-engineering tasks, and production-control tasks were also associated with the group. Initially six assembly engineers and five production-control/materials-handling personnel together with three keymen/supervisors were associated with a group of thirty workers. These indirect operators were located in a work area adjacent to the assembly areas. At a later stage these indirect workers were reduced to seven in total from the previous fourteen, only two assembly engineers and two production-control workers being retained, as much of their work was undertaken by the workers in the area, e.g. method improvements, layout improvements, rebalance, materials handling, etc. This vertical integration of functions had been adopted in order to eliminate unnecessary indirect workers, to give a greater degree of cooperation and involvement, and in order to pass down to the workers on the shop floor, certain of the indirect tasks normally associated with engineers, production controllers and supervisors.

Although certain of the operators within the group were graded differently from any of the others, i.e. associated with higher-level tasks, such operators had in general accepted certain of the lower-level tasks and demarcation between grades had virtually been eliminated. For example, the workers associated with the 'robot' testing, i.e. the electromechanical testing of these machines also undertook to perform certain very simple assembly operations in order to eliminate unnecessary work at previous stations on the line.

Spokesmen were nominated informally by the group, this role being rotated amongst members of the group. The job evaluation employed to fix the pay grade of workers had recently been applied to the various sub-groups as a whole, thus the subgroups were evaluated on performance, quality, etc., rather than individuals, despite the fact that the evaluation scheme was applied to individuals as a routine matter throughout the rest of the company.

Each worker in the group orders his own parts and stock and there is direct worker contact between workers and indirects in other areas such as stores. The manager of the area indicated that little of his time was devoted to routine problems of the line; he occasionally walked around the area, but was rarely called upon to sort out problems in the area.

CASE N   ASSEMBLY OF TELEVISION CONTROL UNIT (HOLLAND)

The television control unit is assembled largely individually by female workers. The basic operations required in the assembly of the item are listed in Exhibit N.1 whilst the arrangement of the work area is shown in Exhibit N.2.

Operation	Description	Operator	Equipment
(1)	Glue protective cover to printed circuit board	Female assembler	Handpress and jig
(2)	Insert bushes into hole in printed-circuit board	Female assembler	Jig
(3)	Insert components	Female assembler	Jig
(4)	Wire crop and solder	Male	Cropping machine, flux and dip solder machine
(5)	Mount contact springs	Female assembler	Jig
(6)	Mount push buttons, etc.	Female assembler	Jig
(7)	Fit assembly into case	Female assembler	
(8)	Test	Male	Test equipment
(9)	Pack	Female packer	

N.B. Cycle time 14 minutes.

Exhibit N.1 Assembly of television controller

All six assembly operations (1, 2, 3, 5, 6, 7) are undertaken by all female assemblers in the group. Each assembler begins to prepare a circuit board at operation (1), moves to the adjacent work-place to finish preparation before moving to one of the other benches in this area in order to insert approximately thirty components into the board. The board complete with jig is then sent to the wire crop, flux and solder area on a moving-belt conveyor, where the male operator on completion of these operations inserts the soldered board in one of several portable racks—each rack being reserved for one assembler. The assembler is required to carry boards in this rack to the other work areas in

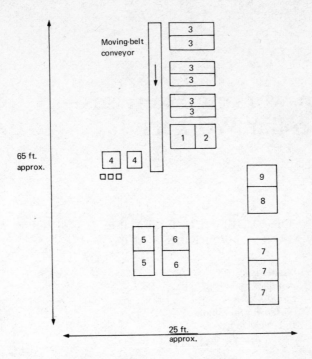

Exhibit N.2 Equipment layout for television-controller assembly.

order to perform the remaining operations prior to testing and packing by other workers.

The total cycle time for one item is approximately 14 minutes of which approximately 12 minutes is undertaken by the assembler.

Twelve to fourteen assemblers work in this area. Together they are responsible for developing a procedure which avoids delay due to waiting for access to a particular work area, which in turn influences their choice of how many items to work on at each operation. They are also responsible to some degree for determining their own output, as they work on a day-rate payment system in which the pay rate is fixed as a result of job grading and in no way reflects output.

References

1. Swedish Employers Federation report by Jan-Peder Norstedt, Stephan Aguren, Feb. 1973.

Appendix C

Group-working Exercises—
Blue-collar Workers

SUMMARY OF PUBLISHED EXERCISES IN GROUP WORKING
AMONGST MANUAL WORKERS

    (1) *Organization*
    (2)   *Reference*
    (3)   *Original Job*
    (4)   *Reason for Change*
    (5)   *New Job*
    (6)   *Effects, etc.*

 (1) (1) Pye Ltd., Lowestoft, U.K.
     (2)   *Manufacturing Management* (December 1972)
     (3)   Production lines for assembly of televisions.
           High speed, short cycle
     (4)   Increased variety for operator.
           Change from work process orientation to product orientation.
           Move to operator.
           To improve quality
     (5)   Teams of seven or eight workers.
           Responsible for quality.
           (1) Casing (tube, loudspeaker and subassembly into cabinet), testing and packing.
           (2) Coils for transformers made by one instead of three operators
     (6)   Cut labour turnover and rejects.
           More flexible production set up.
           Training is simpler—does not disrupt large line.
           Absenteeism and lateness has less effect.
           Reorganization along product rather than process lines makes supervision easier and lines of communication shorter.
           Labour turnover down from 30 per cent to 20 per cent.
           Increased space requirements and extra capital cost

 (2) (1) Philips, Holland
     (2)   Van Beek, H. G. (1964), also Philips Report (1969)
     (3)   104-station line; inspectors placed at end of line
     (4)   Problem of material organization
     (5)   Division of line into 5 groups with buffers between (1 hour).
           Inspectors placed at end of each group

(6)   Reduction of waiting time to 55 per cent.
Increased earnings.
Reduced absenteeism.
System losses reduced.
Quality improved—quicker feedback

(3) (1) Saab, Sweden
     (2)   Palmer, D. (1972) and Holmelius, K. (1972)
     (5)   7 work areas; 3 fitters in each team.
Assemble engine consisting of 90 parts and total work content of 30 minutes either jointly or individually or both.
Ultimately 5 or 6 people at each station?
Women workers as well as males.
Workers can attend Production Group meetings to discuss methods, safety, working environment
     (6)   More space and more tools

(4) (1) Volvo, Kalmar, Sweden
     (2)   McGarvey, P. (1973) and other sources
     (3)   Automobile assembly (new plant)
     (5)   Groups or workers assemble larger part of car.
(Plant to be built 1974)

(5) (1) Norsk Hydro, Norway
     (2)   Gulowsen, J., Hang, O., and Tysland. T. (1969)
     (3)   Processing plant
     (5)   Organized workers in small groups with in process sections with natural boundaries.
Added responsibilities—maintenance, housekeeping.
Operators given further training in maintenance.
Induction course of 200 hours in chemistry, etc.
20 workers in all
Group incentive scheme based upon quantity produced, cost of materials and working hours
     (6)   Operators able to help each other and learn from each other

(6) (1) Donnelly Mirrors, U.S.A.
     (2)   Donnelly, J. F. (1971) and Gooding, J. (1970).
     (5)   Organization of labour force into interlocking work teams.
Boss and his subordinates are required to work together as a group insofar as their work is related.
Key planning decisions and problem solving is done by the whole team.
Leader's role (formerly 'boss') is to support and facilitate the work of his team, not to order and judge it.
Work teams develop plans for cost reduction to offset the pay increases which they seek
     (6)   Absenteeism and timekeeping improved.
Quality improved.
Number of inspectors reduced.
Scrap loss down

(7) (1) Plastic Pharmaceutical Appliances, U.S.A.
     (2)   Davis, L. E., and Canter, R. (1956).

(3)   Assembly line. 29 women workers, each performing 1 of 9 operations. Job Rotation between hard and easy stations every 2 hours. Paced work (?). Moving belt (?)

(4)   Experiment

(5)   *Group Job.* Eliminate conveyor. Workers rotate amongst 9 stations, using batch assembly method.

Individual Job. Individual performs all 9 operations and inspection, also securing materials at individual stations

(6)   *Group Job.* Productivity index fell to 89 average (100 = line). Defects fell from 0·72 to 0·49 per cent average per lot.

Individual Job. Productivity rose slightly. Defects fall from 0·72 to 0·18 per cent average per lot. Increased flexibility of production. Permitted identification of defects. Reduced external service and control function. Developed more favourable attitude towards responsibility, work rate, etc

(8) (1) Textile weaving, India

(2)   Rice, A. K. (1958)

(3)   Twelve specialists to operate equipment assigned to 240 looms in room, i.e.

Weavers	30 looms
Battery filler	50 looms
Smash hand	70 looms
Grater	
Cloth carrier	112 looms
Jobber and assistant	
Bobbin carrier	
Felter motion fitter	
Oiler	224 looms
Sweeper	
Humidification fitter	

Tasks all highly interdependent. Coordination necessary; however, interaction difficult because of allocation pattern

(4)   Mill failed to produce satisfactory productivity and quality levels after intensive study by engineers of layout and work allocation.

Study revealed poor consequences of job designs which centred on worker/machine allocations

(5)   Study travel and communication patterns.

Work groups formed so that single group responsible for operation and maintenance of specific bank of looms, i.e. geographical rather than functional division

(6)   Efficiency rose from average 80 per cent to 95 per cent.

Damage dropped mean 32 per cent to 20 per cent

(9) (1) Coal mining, U.K.

(2)   Trist, E. L., Higgin, Murray, and Pollak (1963) and Trist, E. L., and Bamforth, K. W. (1951)

(3)   *Conventional Longwall System*

Working cycle consists of:

(1) Prepare area for extraction.

(2) Dig out coal.

(3) Remove coal.

Cycle activities divided into 7 specific tasks, each carried out by different task group. Each task to be completed in sequence on schedule over 3 shifts.

Incentive pay for each worker

(4)   Partial mechanization eliminated previous teams of two manual workers. New method (Conventional Longwall) lacked coordination and control. Work cycle often impeded by failure of completion of task in previous shift. Hostility and conflict amongst workers and between workers and management.

High absenteeism. Management spend lot of time on breakdowns, management emergencies, etc.

(5)   *Composite Longwall Method*

Goals set for completion of all tasks in cycle.

Equal earnings. Each shift picked up where other left off and started new cycle if necessary.

Group allocated tasks internally.

Own systems development for rotating tasks and shifts

(6)   Interchangeability of workers developed.

Absence down 20 per cent to 8·2 per cent.

State of cycle progress at end of shift improved.

Productivity improved

(10) (1) Non-Linear Systems Inc., U.S.A. (precision electronic instruments)

(2)   Kuriloff, A. H. (1963)

(3)   Assembly lines. Male workers

(5)   Seven-man teams. Each team headed by technician (Assistant Assembly Manager).

Teams built complete instruments from kits of parts, i.e. insert components into boards, solder, make wiring harnesses, bolt hardware.

Complete subassemblies. Test. Run-in and calibrate instruments.

No formal planning. Team agree work allocation. Pace themselves

(6)   Productivity (man hours/instrument) increased.

Some internal training and development of labour.

Quality improvement (70 per cent complaints)

(11a) (1) Philips—Assembly

(2)   Philips report (1969)

(3)   3 foremen.

15 chargehands.

231 assembly workers and inspector.

Mainly women workers.

Cycle time 5–120 seconds.

All had own job. Foremen responsible for arranging work and individual quality feedback.

Wage system fixed. Individual pay plus performance allowance

(5)   Formation of independent groups. Responsibility for job allocation, supply of materials, quality inspection within the group. Submission to final inspection and appointment of delegates to talk with department management.

Groups worked on a group job.

Job of foreman eliminated. Chargehands given jobs as group leaders with more extensive responsibility, i.e. consultation and assessment.

Number of random sample inspections decreased.

Finally 14 group leaders and 282 assembly workers and inspectors

(6)   Wastage and repairs decreased 7 per cent to 3 per cent as a result of faster feedback of faults.

Pattern of communication with other departments changed as result of deverticalization.

More appeals for solution of problems to ancillary departments.

More satisfaction for workers.

Workers physically fitter, less harassed, but more negative view as to throughflow of work.

Attitudes to colleagues improved

(11b) (1) Philips—electrical subassemblies

(2)   Philips report (1969)

(3)   Assembly lines of 22 or 11 workers. Cycle time 50 seconds.

Pre-processing of materials, electrical adjustment, final inspection and packing, done individually.

Department head, 1 foreman, 4 chargehands each responsible for approximately 20 people

(4)   Reduction in output requirements

(5)   (Individual assembly attempted—abandoned due to technical difficulties.)

Assembly by group of 3 workers.

Cycle time increased 50 seconds to 13 minutes.

Individual quality feedback. Improved layout

(6)   Output up 3 per cent.

Quality improved.

Turnover and absenteeism down.

More attention to training required.

Supervision trained in planning and control for small groups, supervision of stocks, etc.

Need to adapt management structure.

Part of chargehand's job disappeared, i.e. disciplinary tasks.

Greater respect for quality with workers.

Chargehand and foreman's job integrated—'assembly boss'—to manage 30 workers—complex task

(12) (1) G.E.C., U.S.A.—welding work on heavy electronic systems— jobshop

(2)   *G.E. Personnel Research Bulletin*, No. 10 (Jan. 1971)

(3)   12 male operators performing welding and related operations, e.g. drilling and straightening.

Incentive pay

(4)   To reduce operating costs

(5)   Group given almost complete responsibility for all planning, scheduling and control functions in addition to original tasks. Take over many of these tasks from specialists, e.g. Production Control etc.

Receive drawings and specifications, plan the job, sketch the tooling and fixtures, and communicate directly with managers and others as necessary.

Foreman designated 'welding consultant'—leadership from within group.

Paid average earnings until new system fully implemented

(6)   Some initial problems, especially in relations of group with specialist staff.
Improved output.
Improved weld quality.
50 per cent saving in shop overheads

(13) (1) G.E.C., U.S.A.
      (2)   Sorcher, M., and Meyer, H. H. (1968) and Sorcher, H. (1969)
      (3)   Female assembly workers.
                Six work groups typical of assembly-line work
      (5)   Discussion meetings with foremen.
                Tour of manufacturing operation.
                Groups asked to set own quality goals to discuss how quality might be improved
      (6)   Improved productive motivation.
                Attitudes towards supervision improved.
                Improvement in workmanship and quality

(14) (1) Electrical equipment manufacturer, U.S.A.
      (2)   Sirota, D., and Wolfson, A. D. (1972)
      (3)   Assembly of power supplies on highly fragmented assembly line
      (5)   3–5-man teams set up to build entire unit.
                Team decides who should do what.
                Team conducts own quality audit
      (6)   Quantity and quality improved significantly.
                Improved ability to adapt to schedule changes.
                Reduced incidence of part shortages.
                Improved employee attitudes.
                Reduced absenteeism

(15) (1) R. G. Barry Corporation, Ohio, U.S.A.
      (2)   Gooding, J. (1970)
      (4)   Need to expand

(16) (1) Building Materials Company, Sweden
      (2)   Wilkinson, A. (1970)
      (5)   Group of 5 operate production process almost independently of external control, direction and assistance.
                Multi-skilled craftsmen and non-craftsmen carry out maintenance work and report

(17) (1) Paper mill, Norway—chemical pulp department
      (2)   Engelstad, P. H. (1970)
      (5)   Group system, 10–11 people to provide greater responsibility for operation of department and increased understanding and control of processes.
                Specification of group boundaries.
                Definition of control variables.
                Incentive bonus.
                Operators trained for all tasks.

      Special repairman allocated to group.

      Information centre set up.

      Provision for workers to meet together.

      Telephones in work area.

      Group representative for each shift

    (6)   Improved pulp quality

(18) (1) Logging, Norway

      (2)   Gulowsen, J. (1971)

      (4)   Introduction of mechanized equipment

      (5)   5 man team: 3 loggers to fell trees and remove branches;

                           1 driver—logs to production area;

                           1 cutter—cut logs to length.

         Team responsible to inspector and guard.

         Joint piece-rate payment negotiated between management and team.

         No formal or informal leader in team.

         Almost completely autonomous group in respect of work method, working hours, absence, work times, etc.

         Group approved all new members

(19) (1) Lime works, Norway

      (2)   Gulowsen, J. (1971)

      (4)   No change

      (5)   Two work groups of approximately 10 workers.

         Joint piecework for group.

         Production planning for day determined in meeting between manager and group—also decided amount of maintenance and cleaning.

         Groups have complete control of method of working and division of tasks between members.

         Responsible for own recruitment

(20) (1) Christiana Spigerverk, Norway—manufacture of rail springs

      (2)   Gulowsen, J. (1971)

      (4)   No recent change

      (5)   Group of 4: 2 operate punch press;

                           1 operates cutting machine;

                           1 inspects and packs.

         Frequent breakdowns required some flexibility of working, and required the occasional (20–50 per cent of his time) use of one repair man.

         One group member each week responsible for calling repair man.

         Workers formed own work pattern and adjusted to breakdown, minor schedule changes, etc.

         Group bonus

(21) (1) Coal mine, U.K.

      (2)   Herbst, P. H. (1962)

      (4)   Experiment to test new work method (Composite Longwall method—see above).

      (5)   8-man group with complete responsibility for cycle of operations, i.e. deployment of members.

         Group pay

(22) (1) Sack manufacturer, Sweden
  (2) Wilkinson, A. (1970)
  (3) Machine minding
  (4)   Increased manning led to lower productivity
  (5)   Group given task of making good rejects and deciding output.
     New payment system
  (6)   Increased output.
     Reduced wastage

(23) (1) ICI Ltd., U.K.
  (2) Paul, W. J., and Robertson, K. B. (1970)
  (3) Process operators
  (5)   Group of 16 operators given responsibility for efficiency and initiating maintenance.
     Able to fix own work breaks and organize own cover.
     Some involvement in decision making
  (6)   Difficulty in persuading operators to accept responsibility.
     Possibly some increase in output

(24) (1) ICI Ltd., U.K.
  (2) Cotgrove, S. *et al.* (1971)
  (3) Nylon spinning; 40 employees
  (4)   Productivity deal
  (5)   Plan and organize own work, meal breaks and rest pauses.
     Responsibility for scheduling, quality inspection and machine checks.
     Preceded by meeting to produce ideas
  (6)   Great productivity.
     Fewer faults.
     Fewer grievances

(25) (1) Texas Instruments, U.S.A.
  (2) Weed, D., in Maher, J. R. (1971)
  (3) Factory and office cleaners
  (4)   Poor standard of work of contractors
  (5)   Groups given specific areas of responsibility.
     Plan and control own work.
     Training in work simplification
  (6)   Cleanliness improved.
     71 personnel used instead of 120.
     Turnover reduced substantially.
     Saving $103,000 p.a.

(26) (1) Shell UK Ltd.—Stanlow refinery
  (2) Taylor, L. K. (1972)
  (3) Plant operators in micro-wax department
  (5)   Team problem solving.
     No clocking in. Decide own shift rota.
     Increased job content. Increased training and promotional prospects
  (6)   Cost of testing reduced

(27) (1) Volkswagen, Germany
    (2)   Taylor, L. K. (1972)
    (3)   Assembly-line work
    (5)   Groups held responsible for quality.
                  Participation in decision making.
                  Work on whole subassemblies.
                  Requisition own materials and equipment.
                  Group budgeting.
                  Authority to disgard faulty materials.
                  Draw up schedules. Initiate investigation

(28) (1) Hellerman Deutsch
    (2)   Taylor, L. K. (1972)
    (3)   Manufacture of high-precision components
    (5)   Participation in making improvements.
                  Training to make worker more flexible.
                  High degree of self-supervision

(29) (1) Volvo, Sweden
    (2)   Palmer, D. (1972)
    (3)   Upholstery department. Assembly workers
    (5)   Participation in decision making and elimination of foremen

(30) (1) General Foods, U.S.A.
    (2)   *Business Week* (9 Sept. 1972)
    (3)   Production workers
    (5)   Assignment of tasks to teams of 7 to 17 members.
                  Workers learn each job.
                  Each team covers entire phase of operation from processing raw material to end-product, packaging, shipping and office work.
                  Time clocks removed.
                  No conventional departments
    (6)   Quality improved

(31) (1) T.R.W., U.S.A.
    (2)   *Business Week* (9 Sept. 1972)
    (3)   Assembly-line work
    (5)   Group assembles product scheduling own work

(32) (1) Company B, U.S.A.
    (2)   Skinner, W. (1971)
    (3)   General-purpose job shop
    (4)   Low labour productivity
    (5)   Set own rules and regulations to achieve required output. Check quality schedule.
                  'Trainer' available to help.
                  30 workers involved

(33) (1) Harwood Manufacturing Co., U.S.A.
    (2)   Viteles (1953), and Coch, L., and French, J. R. P. (1948)
    (3)   Clothing manufacture.
                  Garment manufacture.
                  Groups of female workers
    (4)   Resistance to change of job and work methods

(5)   3 experimental groups participated in decision making in connection with job (and associated piece rate) changes, and in the planning of such changes.

4th group retained as control group.

Changes made without participation

(6)   Output of control group fell, and remained low, following change. Marked resistance.

Output of each of 3 experimental groups improved.

No signs of resistance

(34) (1) Harwood Manufacturing Co., U.S.A.

(2)   Viteles (1953) and French, J. R. P. (1950)

(3)   Clothing manufacture.

Sewing-machine operators.

Groups of from 4–12 female workers

(5)   Teams asked to consider setting group output goals, and deciding time by which goals might be achieved

(6)   Output of experimental groups in comparison to control groups increased 18 per cent (average)

(35) (1) Hovey and Beard Co., U.S.A.

(2)   Strauss (1955)

(3)   Wooden toy manufacture.

8 female trainee workers employed on painting operations.

Flow-line work, items removed from moving belt and replaced after spraying.

Group bonus

(4)   Absenteeism and turnover.

Complaints about working conditions.

Poor learning progress

(5)   Meetings of workgroup and foreman to discuss working conditions.*

Further meeting on speed of line; workers ask to control line speed†

(6)   * Some changes in ventilation made.

† Average belt speed increased; fewer items 'missed'; quality improved.

Output and earnings increased beyond scheduled target.

N.B. Original situation restored following complaints of inequity from elsewhere in plant

(36) (1) Serck Audco Ltd., England

(2)   *Production Planning and Control*, NEDO, 1966

(3)   Engineering batch production of a range of valves.

Conventional layout by process

(4)   To improve delivery performance and throughput time.

To reduce work in progress. To rationalize product line

(5)   Group technology system introduced in which 'flow-line' cells of machines and workers were created. Each cell consists of up to 15 simple machine tools and 6 or 7 operators, with the machines linked by conveyors. Equipment in cells is devoted to production of one type of component. Operators required to be flexible. Team in cell is controlled by working supervisor and together determine

how to process work load specified for period. Some movement of workers between cells. Attendant changes in production planning and data processing systems

(6)   Work in progress reduced. Throughput time down. Late deliveries reduced

(37) (1) Footwear manufacturer, Norway
       (2)   French, J. R. P., *et al.* (1960)
       (3)   Assembly of footwear by groups of four workers (males and females)
       (4)   Experiment (replication of Coch and French (1953))
       (5)   4 control groups and 5 experimental groups.
                  Experimental groups participate, through group discussion with management, in decisions relating to introduction of new models, i.e. allocation of articles to group; length of training; division of labour and job assignments in group
       (6)   No difference between experimental and control groups in respect of output. Participation was thought to be related to resistance to change and improved worker/management relationships in certain cases

(38) (1) Electrical equipment manufacturer
       (2)   Huse, E. F., and Beer, M. (1971)
       (3)   Forming of tubes and other shapes for electrodes. Females operating lathes
       (4)   O.D. programme
       (5)   Teams of workers (4?). Given responsibility for certain products, i.e. allocation of tasks and establishing schedules against output targets
       (6)   Not successfully introduced initially, but later more effective approach to same objective adopted.
                  Improvement in productivity (20 per cent).
                  Improvement in involvement and commitment of workers

(39) (1) Banner Company, U.S.A.
       (2)   Bowers, D. G., and Seashore, S. E., Evan, W. M. (1971)
       (3)   3 of 5 departments in plant manufacturing packaging materials. Manufacturing involves use of high-speed semi-automatic equipment. Male and female workers
       (4)   Experiment
       (5)   Changes—increased emphasis on work group rather than individual worker as the unit supervised; higher rate of interaction in groups, higher degree of participation in decision making and control activities in lower echelons of organization, more supportive behaviour by supervisors
       (6)   Changes tended to lead to improvements in employees' attitudes to work and conditions of employment

(40) (1) Ford, U.S.A.
       (2)   *Life* (September 1972)
       (3)   Motor-car instrument-panel assembly
       (4)   Experiments
       (5)   Team assembly

(41) (1) Chrysler, U.S.A.
    (2)   *Life* (September 1972)
    (3)   Motor-vehicle-engine assembly
    (5)   Team invited to hire own people and set up own production line
    (6)   Absence down

(42) (1) Motorola, U.S.A.
    (2)   Sirota, D., and Wolfson, A. (1972)
    (3)   Machine workers
    (5)   Elimination of inspectors. Groups used for decision making and also work on complete modules

(43) (1) Hoover Ltd., U.K.
    (2)   *Management Today* (May 1973)
    (5)   Work teams with job rotation facility and group payment system

(44) (1) John Player and Sons, U.K.
    (2)   Hallam, P. A. (1973)
    (3)   Warehouse work
    (4)   To provide flexibility, balance work loads
    (5)   Group of 12 men.
              Reduction of demarcation between workers.
              Job rotation.
              Workers given extra responsibility for clerical work, etc.
              Extra pay commensurate with increased job knowledge
    (6)   Improved satisfaction.
              Better communications.
              Increased flexibility

(45) (1) H. P. Hood, U.S.A.
    (2)   Foulkes, F. K. (1969)
    (5)   Workers watched film of work method and helped determine new layout in order to improve efficiency

References

Bowers, D. G., and Seashore, S. E., 'Changing the structure and functioning of an organisation' in Evan, W. M. (Ed.), *Organisational Experiments: Laboratory and Field Research*, Harper and Row, New York, 1971, pp. 185–201.

Business Week, 'Job monotony becomes critical', *Business Week* (9 September 1972); 'Management itself holds the key', *Business Week* (9 December 1972).

Coch, L., and French, J. R. P., Jr., 'Overcoming resistance to change', *Human Relations*, **4**, 1 (1948), pp. 512–533.

Cotgrove, S., Dunham, J., and Vamplew, C., *The Nylon Spinners: A Case Study in Productivity Bargaining and Job Enlargement*, Allen and Unwin, London, 1971.

Davis, L. E., and Canter, R., 'Job design research', *Journal of Industrial Engineering*, **VII**, 1 (1956), p. 3.

Donnelly, J. F., 'Increasing productivity by involving people in their total job', *Personnel Administrator*, **34**, 5 (September–October 1971).

Dyson, B., 'Hoover's group Therapy', *Management Today* (May 1973).

220

Engelstad, P. H., 'Socio-technical approach to problems of process control', in Davis, L. E., and Taylor, J. C. (Eds.), 'Design of Jobs', Penguin Books Ltd., London. 1970.

Evan, W. M. (Ed.), *Organizational Experiments: Laboratory and Field Research*, Harper and Row, New York, 1971.

Foulkes, F. K., *Creating More Meaningful Work*, American Management Association, New York, 1969.

French, J. R. P., Jr., 'Field experiments: changing group productivity', in Miller, J. G. (Ed.), *Experiments in Social Process: A Symposium on Social Psychology*, McGraw-Hill, New York, 1950.

French, J. R. P., Israel, J., and Aas, D., 'An experiment on participation in a Norwegian factory', *Human Relations* **13**, (1960), pp. 3–19.

G.E.C., *G.E. Personnel Research Bulletin No. 10*, U.S. G.E.C., 1971.

Gooding, J., 'It pays to wake up to the blue collar worker', *Fortune*, **LXXXII**, 3 (September 1970), pp. 133–135.

Gulowsen, J., Hang, O., and Tysland, T., 'Norwegian firm taps total human resources', *Industrial Engineering*, **1**, 8 (1969), pp. 30–34.

Gulowsen, J., 'A measure of work group autonomy', in. Davis, L. E., and Taylor, J. C. (Eds.), *Design of Jobs*, Penguin, 1972.

Hallam, P. A., 'An experiment in group working', *Work Study and Management Services*, **17**, 4 (April 1973), pp. 240–244.

Herbst, P. H., *Autonomous Group Functioning: An Exploration in Behaviour Theory and Management*, Tavistock Publications, London, 1962.

Holmelius, K., 'Time in factories—The end of the great timing machine, the production line', paper to 'Time at Work' Conf., London, (12 October 1972).

Huse, E. F., and Beer, M., 1971, 'Eclectic approach to organisation development', *Harvard Business Review*, **49** (September/October 1971), pp. 103–112.

Kuriloff, A. H., 'An experiment in management—putting theory Y to the test', *Personnel*, **40**, 6 (November–December 1963), pp. 8–17.

Life, Sept. 1972.

McGarvey, P., 'Car making without the boredom', *Sunday Telegraph*, (4 February 1973).

Maher, J. R., *New Perspectives in Job Enrichment*, Van Nostrand Reinhold, 1971.

Manufacturing Management, 'Work teams at Pye beat production line problems', *Manufacturing Management* (December 1972), pp. 18–19.

N.E.D.O., *Production Planning and Control*, H.M.S.O., London, 1966.

Palmer, D., 'Saab axes the assembly line', *Financial Times* (Tuesday May 23 1972).

Paul, W. J., and Robertson, K. B., *Learning from Job Enrichment*, ICI Ltd., Central Personnel Dept., London, 1970.

Philips Report, *Work Structuring: A Survey of Experiments At NV Philips, Eindhoven 1963–8*, NV Philips, Eindhoven, 1969.

Rice, A. K., *Productivity in Social Organisation—The Ahmedabad Experiment*, Tavistock Publications, London, 1958.

Sirota, D., and Wolfson, A. D., 'Job enrichment: what are the obstacles', *Personnel*, **49**, 3 (May–June 1972), pp. 8–17; 'Job enrichment: surmounting the obstacles', *Personnel*, **49**, 4 (July–August 1972), pp. 8–19.

Skinner, W., 'The anachronistic factory', *Harvard Business Review*, **49** (January–February 1971), pp. 61–70.

Sorcher, M., and Meyer, H. H., 'Motivating Factory employees', *Personnel*, **45**, 1 (January–February 1968), pp. 22–28.

Taylor, L. K., *Not For Bread Alone—An Appreciation of Job Enrichment*, Business Books Ltd., London, 1972.

Trist, E. L., and Bamforth, K. W., 'Some social and psychological consequences of the Longwall method of goal getting', *Human Relations*, **IV**, 1 (February 1951), pp. 3–38.

Trist, E. L., Higgin, G. W., Murray, H., and Pollack, A. B., *Organizational Choice*, Tavistock, 1963.

Van Beek, H. G., 'The influence of assembly line organisation on output, quality and morale', *Occupational Psychology*, **38**, 3 and 4 (July and October 1964), pp. 161–172.

Weed, D., 'Job enrichment "clean-up" at Texas Instruments', in Maher, J. R. (Ed.), *New Perspectives in Job Enrichment*, Van Nostrand Reinhold, New York, 1971, pp. 55–77.

Whyte, W. F., *Money and Motivation*, Harper and Row, New York, 1955.

Wilkinson, A., *A Survey of Some Western European Experiments in Motivation*, Institute of Work Study Practitioners, London, 1970.

Name Index

224

Subject Index